Cogeneration Power Plants

COGENERATION POWER PLANTS

PLANNING AND EVALUATION

JOEL K. WILSON

PennWell®

Disclaimer

The recommendations, advice, descriptions, and methods in this book are presented solely for educational purposes. The author and publisher assume no liability whatsoever for any loss or damage that results from the use of any of the material in this book. Use of the material in this book is solely at the risk of the user.

PennWell Corporation
1421 South Sheridan Road
Tulsa, Oklahoma 74112-6600 USA

800.752.9764
+1.918.831.9421
sales@pennwell.com
www.pennwellbooks.com
www.pennwell.com

Publisher: Matthew Dresher

Library of Congress Cataloging-in-Publication Data

Names: Wilson, Joel, 1961- author.
Title: Cogeneration power plants : planning and evaluation / Joel Wilson.
Description: Tulsa, Oklahoma, USA : PennWell Corporation, [2018]
Identifiers: LCCN 2018049631 (print) | LCCN 2018050837 (ebook) | ISBN 9781593706470 (ebook) | ISBN 1593704178 (hardcover)
Subjects: LCSH: Cogeneration of electric power and heat.
Classification: LCC TK1041 (ebook) | LCC TK1041 .W57 2018 (print) | DDC 333.793—dc23

Printed in the United States of America

1 2 3 4 5 23 22 21 20 19

This book is dedicated to my beloved wife, Corinne, whose help with this project was invaluable. I will cherish her love and support as long as I live.

Contents

Figures

Tables

1

Introduction

There are many good reasons to want your own power plant, including improved power quality (voltage and frequency), increased reliability, and savings on energy expenses. This book is intended to help those interested in cogeneration power plants by laying out a thorough and proven planning methodology for new facilities, as well as an evaluation methodology for existing facilities. The main difference between these two methodologies is that, with an existing facility, because all of the electrical and thermal loads are already in place, you can examine how and when they are used. By contrast, with a new facility, one designed but not yet built, the electrical and thermal loads will be identified (and perhaps even specified), but their operational nature will be an estimate at best (and a guess at worst).

Cogeneration, also known as *combined heat and power* (CHP), is the generation of more than one kind of energy from a single fuel source. The typical energies produced using a wide variety of fuels are electricity and thermal (usually heat). Any industrial or commercial facility that is changing the temperature of something—and it doesn't really matter what—is a candidate for the application of CHP. Turning water into steam, hot water, chilled water, or chilled air, for example, is commonly done to enable the processes of industries, from making cardboard to frozen peaches (and everything in between). Cogeneration is the most efficient solution for *distributed generation,* which is becoming an increasingly popular choice, rather than relying on the centralized power distribution structure we've been accustomed to for decades. That is, in addition to (or instead of) using electricity from the local electric utility via the existing power grid, many industrial and commercial facilities are choosing to install their own on-site power plant. These are smaller power plants than the utility-scale units supplying the grid and are typically under 10 megawatts (MW) in size. They harvest the leftover heat from the combustion process—heat that used to be wasted—to make a thermal component, usually hot water, steam, and/or chilled water. Conventional large-scale utility power plants can operate at efficiencies around 35%. On-site cogeneration power plants can easily operate at efficiencies of 80%—and even higher when using water vapor recovery equipment.

When I started my third decade in the power generation industry in 2005, I made the shift from nuclear power to conventional cogeneration. Over the course of the next 11 years, I saw over a dozen CHP (cogeneration) power plants either fail outright or perform far below expectations—all because of poor planning and evaluation of the power plants from the beginning. Poor planning with regard to engineering, construction, and/or operations and maintenance often brings companies developing and selling cogeneration power plants to their knees—and in many cases to the point of collapse—as a result of the combination of wishful thinking, intentional oversight, and a failure to grasp the obvious.

If you put enough torque on something, you can spin it however you want to. If you are trying to sell engine-generator sets or even entire cogeneration packages, you can paint a rosy picture for your potential customer by using the *nameplate data* (the maximum rated output) of your electrical generators and your heat recovery system. The reality, though, is that these systems will not perform at their full rating on an annual basis unless they are sized to provide some kind of minimal baseload only, such that the electrical generation equipment is fully loaded at all times. For example, if you were to buy a new car because it gets great gas mileage while driving down the highway and expected that level of fuel efficiency for all your driving conditions, then you would be disappointed at the amount of money you'd spend on fuel every month. This is similar to the situation where a customer installs a photovoltaic power system of a given size expecting that the kilowatt (kW) rating of the panels, multiplied by the number of hours of daylight, will yield a daily, expected, total generated output from their solar power plant. Unfortunately, such a short-sighted customer will soon have to embrace the reality that the sun doesn't always shine and that a single passing cloud can eclipse the sun and cause the power output to plummet. On an annual basis, the overall performance of *any* power plant will not be even close to expectations based on nameplate data alone. Any forecasts made at the beginning of a power plant project must be conjured using a heavy dose of reality to avoid such disappointments, and my methods will make your projections more realistic than they may be now.

This book covers a wide range of power and heat sources—not just reciprocating engines and combustion turbines—and examines possible hybrid power plants that also use other technologies, including renewable energy sources. I am presenting what I have learned about developing and analyzing cogeneration power plants; that is, I'm not going to tell you how you should go about your analysis, but rather describe how I do my analysis. Applying the methods I use to your given situation will give you a realistic assessment (or prediction) of your power plant's operational and financial performance. You don't have to be a rocket scientist to come up with your own models to figure out what the real performance of your cogeneration power plant will be, in terms of both the operation of the power plant and the financial side of things, using a year's worth of utility load data and the rate structure of the given utility for the customer at hand. I describe the models I use in chapter 6, and explain why I analyze the data in particular ways, so that you can incorporate as much of my methodology into your models as you wish.

One can learn a lot about cooking by reading books and watching the cooking channel, but to become a chef, one must enter the arena of the kitchen and actually cook—and thereby also learn by doing. Similarly, reading this book alone is only part of what you will need in order to understand the complexities associated with a cogeneration power plant; to fully understand, you will need to perform all of the analysis and evaluation for yourself. It is only in the application of the knowledge gained from reading about something and the effort of doing that given something where theory and reality diverge. Once you determine that a facility is viable for cost-effective on-site power, it becomes reasonable to invest the time to perform a detailed analysis of the situation, and this book is intended to help with that as well.

I use a *systems engineering* approach to problem solving. Thus, I begin the book with some background information on systems engineering and why it is a good fit for this application. I then delve into the analysis of the site of the potential power plant and the available historical data of energy consumption. I present a wide range of power and heat sources that are available for use in an on-site power plant (though by no means an exhaustive list) along with the pros and cons associated with each technology. Next I describe the details of the performance models I use to see how well the given design concept actually holds up over a full year of operation, including operational costs. Finally, a financial analysis is done to see if, over the lifetime of the equipment, the design of the power plant makes economic sense. If it does, the design concept can be refined; if not, then an investigation into the data to see where the problem is can help preclude the mistake from being made in future designs.

I also include several case studies of cogeneration power plant successes and failures, each with a particular set of lessons learned. I am a firm believer in learning from the failures of others, and while none of the power plants I describe were developed, constructed, or operated by people intentionally bent on the failure of the projects, the results were clearly not what they wanted. Owning and operating a cogeneration power plant is like being the owner of an old house, in that there's always something that needs attention or repair. It helps if you're already holding a general contractor's license, but you can also learn as you go with a do-it-yourself approach. It also follows that the next time you buy a house, whether it's old or new, you will be that much more alert to what could go wrong in the future.

This book is not intended for the absolute beginner and presumes that the reader has some knowledge of electricity, basic physics, and power plant-related terminology. I attempt to explain the material in a way that opens it up to a broader audience than a collection of engineers; yet, if anything is not clear, the Internet is available for your research. I offer an approach to solving energy management problems; however, I do not include such elaborate details as heat balances because there are too many ways of building power plants to cover them all. Will you be using stainless steel piping or PVC? Schedule 40 or schedule 80? Insulation or lagging on your piping or neither? Those final engineering details remain for you to decide.

2

The Systems Engineering Approach

As Applied to Cogeneration Power Plants

Systems engineering, as a distinct engineering discipline, evolved from needs of the U.S. space program and military in the 1950s and 1960s. With the emergence of electronic devices, from transistors to the first computers, and the increasing complexity of both space-based systems and military equipment, a formal method was needed to coordinate all of the various subsystems into one cohesive whole. The military and space agencies realized that complex systems such as spacecraft were composed of hundreds or even thousands of subsystems and that each of those subsystems was probably going to be developed by a separate contractor. Somehow, when combined, all of those subsystems had to work on their own and play well together for the overall spacecraft or weapon system to operate as designed. This gave birth to the systems engineering program.

Systems engineers are multidiscipline-trained engineers. Systems engineering is not limited to mechanical or electrical systems (hardware) nor to computer software and control systems, but also applies to *how* things are done. For example, when you go to the grocery store and are ready to leave, most people look for the shortest line to get into, so they can pay for their groceries and leave in the least amount of time. When you go to a bank, though, you will typically wait in a single line and then go to the next available teller. Both systems accomplish the same task and involve the same steps, like waiting in a line until you can interact with the teller or cashier, but which system is better? The bank system is the better one, saving the average customer 17%–22% in terms of time wasted while standing in line. This is one reason grocery stores on military bases use the bank method. However, despite the potential process improvement, civilian grocery stores generally wouldn't want to surrender the necessary floor space to accommodate one, longer line (instead preferring to pack more stuff for you to buy into the store).

Basic systems engineering comprises four areas: *functions, requirements, attributes, and tests* (FRAT). These categories are illustrated in figure 2–1.

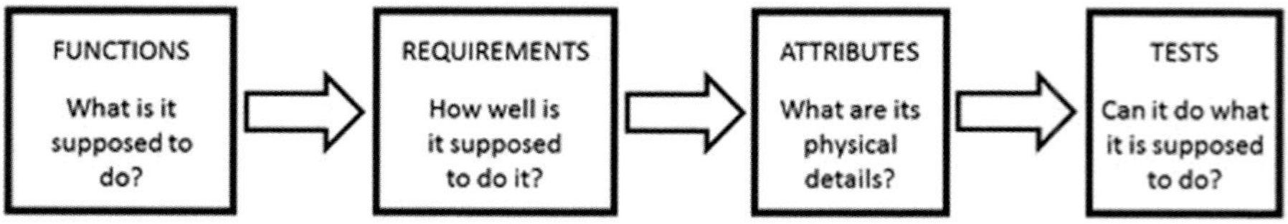

Fig. 2–1. FRAT mnemonic for systems engineering

For the function category, envision starting at the top of very large systems and working your way down. What is the function of the overall system? From there it's divided into subfunctions. Consider the automated teller machine (ATM). For most people, the primary function of an ATM is to dispense cash, but there are other functions available, such as depositing both cash and checks and accessing multiple bank accounts to transfer funds between them. Naturally, each of those functions can be broken down into multiple subfunctions.

For each function, whether on a large scale or one of the minutia, there can be a set of associated requirements that dictate how well the function needs to do what it does—how accurate, how fast, and so forth.

Once the functions and their requirements are identified, work can begin on conceptual designs. There's rarely only one way to meet the objectives, and as designs take shape, certain attributes can be assigned. For a cogeneration power plant, these attributes include physical details, such as using a combustion turbine instead of a reciprocating engine as the prime mover or using some of the steam flow for evaporative wastewater distillation units. Some attributes will be dictated by the equipment used, whereas others can be as personal as the paint color.

The final category involves coming up with ways to test the system to prove whether the functions are performed within the specifications called for under the metrics set forth in the established requirements. The results of these tests are then used to refine the function's (and the requirement's) specifications to further enhance the final product or system in an iterative fashion.

For a cogeneration power plant, the approach could start like this:

1. Determine the specific location of the power plant and any site-specific factors that may possibly affect any potential solution to the stated problem.
2. Identify all the possible heat source solutions to the stated problem.
3. Perform a feasibility analysis of the possible heat source solutions.
4. Refine the analysis of possible heat source solutions.

Life-Cycle Cost and Modeling

When comparing the various possible solutions, always compare each possible solution to the *do-nothing* option as a baseline. The use of models in analysis and evaluation of the performance of a cogeneration power plant design, as well as in load analysis, is critical. Importantly, remember that the results of your analysis and analysis are only as precise as the models used. If your models are imprecise, you will ultimately be making decisions using data based more on wishful thinking than reality.

For example, when modeling engine-generator performance, you must account for your generator's derating as changes in ambient temperature occur. Furthermore, you must account for other factors such as the changes in heat rate, emissions, exhaust temperature and flow, and overall efficiency when the engine is running at less than full load. Two other critical parts of the evaluation of all the possible solutions are as follows:

- Consider all possible solutions. The failure to do so is the main cause of suboptimal performance in cogeneration power plants.
- Evaluate all alternatives, not just the lowest-cost option. Consider the initial costs, long-term costs, the ability to increase size as the customer's size increases, anticipatory requirements (e.g., continuous emissions monitoring), and any other desired bells and whistles.

Systems engineering is used not only to solve engineering problems but also to ensure that the customer's needs are satisfied throughout a system's entire life cycle. This is accomplished using an iterative process composed of the following general steps (fig. 2–2):

1. *Fully understand the problem.* Define the given problem. This is the most important systems engineering task. Identify specific needs, system functions, requirements, and attributes over the course of the life cycle of the project. Awareness of these critical elements will help in finding an optimum solution to the given problem.
2. *Identify possible solutions.* Once the overall system requirements have been determined, thoroughly investigate possible solutions, including comparisons and trade off studies, cost-benefit analysis, risk analysis, and other methods.
3. *Perform detailed analysis.* As the solution is approached, undertake a detailed analysis, including life-cycle modeling of the system. This results in a refinement of system requirements and identifies ways of reducing costs, integrating fragmented processes, identifying communication and control interfaces, and resolving other, previously unforeseen problems.
4. *Complete system testing and performance assessment.* Once a solution to the given problem has been found, thoroughly test it against the metrics of the requirements set. A complete and thorough analysis of

the performance of the system is conducted, and the resulting evaluation verifies and validates the solution—or identifies areas of concern.

5. *Integrate improvements.* Any and all areas where the performance analysis identifies the need for improvements are then incorporated into the next iteration of the systems engineering process, further refining the solution to the given problem. Subsequently, the cycle repeats, from step 2, assuming that the initial conditions have not changed and that the original problem is well understood. In real life, though, the problem can be like a moving target, which is why it is critical to involve all of the various stakeholders for the on-site power plant project in the discussions that determine the requirements of the power plant. (The consequences of not doing this are described in the case studies in chap. 7.)

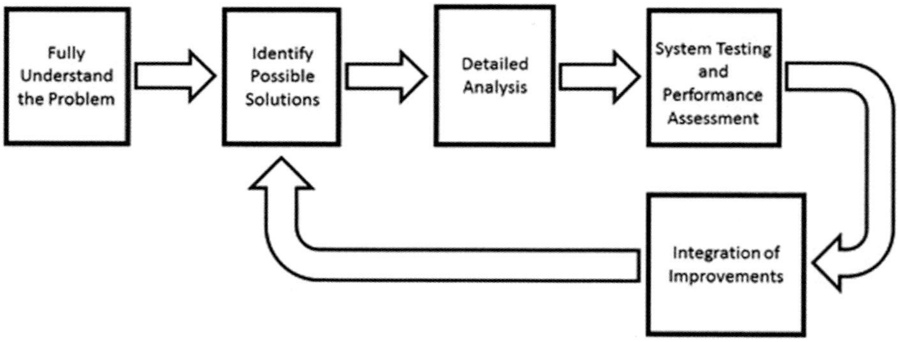

Fig. 2–2. Systems engineering process

The application of systems engineering to the planning and evaluation of a cogeneration power plant is summarized in figure 2–3.

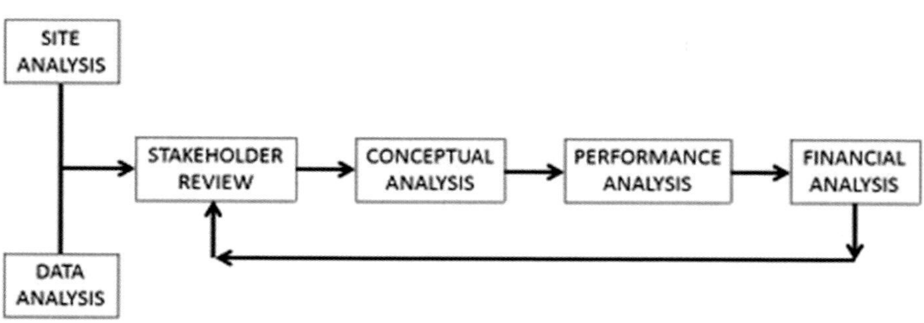

Fig. 2–3. Cogeneration power plant planning block diagram

As will be described in chapter 4, the process starts with evaluating the potential of the site of the power plant both physically and analytically. Once the site is deemed feasible, a detailed conversation needs to take place among all of the stakeholders in the project: the customer, the cogeneration power plant developer, the financial parties, and any others. A big part of this stakeholder meeting is to establish and define the functions and requirements of the power plant, answering questions such as the following:

- How much electricity and thermal energy will the power plant need to make in the first years of operation?
- How much electricity and thermal energy will the power plant need to make two years from now? five years from now? ten years from now?
- Where is the best location to build the power plant?
- What are the concerns of the customer?
- What is the upper limit on the amount of the available funding?
- Are there any regulatory issues to address (wastewater, emissions, etc.)?
- Are there sufficient resources available at the site (water, fuel, etc.)?
- What is the equipment reliability and risk management plan?

Once everyone involved in the development of the power plant project understands the desired capabilities of the plant (as well as the limitations of the plant), ideas can emerge about how to satisfy the defined functions and requirements. It is at this point that the power plant developer usually takes charge. Although the potential customer is always welcome to join the party, most of the people who run such facilities that would benefit from on-site power are not power systems engineers. As the developer's conceptual designs are refined, those original functions and requirements must be referred to after each iteration to ensure that the design doesn't stray from what needs to be achieved. I have been to dozens of power plants where at least one of the stakeholders was unhappy or unsatisfied because the power plant that was built failed to deliver on promises made to them at the start of the project. All of the concerns expressed by each of the stakeholders must be documented at the beginning and then addressed after each iteration of the design process, to avoid 15–20 years (or the life span of the equipment) of frustration and disappointment. Often, this can be difficult, because the needs and desires of the customer are likely to change during the first few months of the project. At some point, a decision must be made about finalizing the requirements, with all parties understanding that future changes to the list will be dealt with as a change order.

Consider the analogy of wanting to build a car: There are a lot of ways to do it, and there's a long list of design features to consider that will affect the performance of the fully-assembled vehicle. Rear-wheel drive or four-wheel drive? Off-road capable or paved streets only? Two doors, four, or five? You get the idea. If everyone agrees up front that it will be an ultraefficient hybrid commuter vehicle, then imagine the frustration on the customer's part when a sport utility vehicle or a snowmobile is delivered instead. For a cogeneration power plant, general options include the following abilities:

- *Black start* (the ability to start the power plant with from electrical blackout conditions)
- *Transient ride-through* (having uninterruptible power supply and control system flexibility to handle large electrical transients)
- *Protection* (isolation and protection from electrical faults on the grid)
- *Thermal deliverables* (steam, hot water, chilled water, hot air, chilled air, etc.)
- *Off-grid* (able to operate independently from the local electric utility grid, as well as water, sewer, and gas pipelines)

For each conceptual design (that meets the functional needs of the facility) a detailed analysis must be done to verify the performance of the power plant over time. Computer models, using software either purchased or developed in-house, can be used to assess the life-cycle performance in all respects: energy output (electrical and thermal), equipment and construction costs, fuel costs, maintenance costs, projected revenue, projected savings over the utility, and so forth. Once the energy needs are proven to be met, the next step is to see if the project makes financial sense—the performance information can be run through a financial pro forma analysis. At the end of the life of the equipment, usually 15–20 years, how will the power plant perform monetarily? Is it financially feasible? Does a sensitivity analysis show any vulnerable areas? Finally, the results of all of the analyses get reported back to the collection of stakeholders as an option and the entire process then goes through another iteration, continuing until all of the possible solutions to the original problem are identified. Again with reference to the car analogy, an economy-class sedan will be as functional as a luxury-class vehicle, but does the customer want to spend the extra money for those luxuries?

All of this sounds a lot more arduous and repetitive than it really is. Taking the time and effort to use this approach will minimize the possibility of buyer's remorse or ending up with a power plant that cannot fulfill promises made to the customer at the beginning. It has been said that the faintest ink is better than the strongest of memories. If proper documentation is kept about what is discussed during the stakeholder meetings at the beginning of the process, especially the agreed list of the proposed power plant's capabilities, then personal expectations can be held accountable against the documented promises made. Additionally, all of the stakeholders must be made aware of the costs, in terms of both the financial impact and the logistics (project schedule) delays that are associated with changes to the power plant configuration after the design is finalized.

3

Considerations for a Cogeneration Facility

Conventional Power Distribution

The electricity available at the electrical outlets in your home or business was generated in a power plant somewhere, typically a very long way from where that electricity is being used. In the standard, historical scheme, the regional electric utility owns and operates large power plants (hundreds of megawatts in size) that send out their power over high-voltage transmission lines. These lines can be hundreds of miles long, since the utility's power plant is not usually close to the cities that consume the electricity. When electricity is needed, power is tapped off these lines via substations, to deliver the electricity at lower voltage levels, to low-voltage transmission lines. Power is again tapped off these lines, via more substations, and the process continues until the power line on your street supplies electricity directly to your home or facility (via a transformer) at the 120 V or 220 V level. A common analogy is to think of a tree, where very-high-voltage transmission lines represent the main branches that then have other, smaller branches emanating from them (lower-voltage transmission lines) and ultimately terminate in individual leaves (consumers).

By contrast, in the distributed generation scheme, electricity is generated close to where it is consumed, if not directly at the site of the consumer. This would be like taking the nutrients supplied by the roots of a tree and delivering it directly to a twig, with several leaves on it, or to an individual leaf—effectively bypassing the typical transmission and distribution infrastructure.

The traditional power generation and distribution system is shown in figure 3–1. Large—1,000 MW and above—remote power plants (that no one wants in their backyard) generate electricity that is then stepped up to very high voltage levels by transformers before going onto the grid. Transmission lines—the type you see from the highway, with their enormous, skeletal support structures—bring that very high voltage power closer to populated areas where substations reduce the voltage from *very high* (765,000 V) to *high* (138,000 V) levels and continue the distribution toward more densely populated areas. Additional substations are used to reduce the voltage again, to about 69,000 to 26,000 V (*medium* voltage), before the process continues to local substations, close to the point of use, which further reduce the voltage to levels on the order of 12,470 V, which flows through the power lines along city streets before being delivered to commercial sites and residential areas.

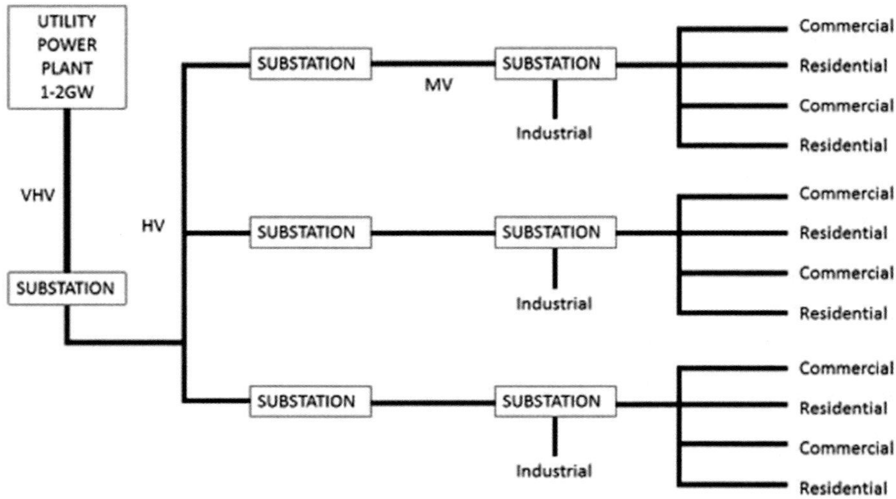

Fig. 3–1. Conventional power distribution

Distributed generation approaches the supply problem from the other way round, being physically close to the point of use (or actually on-site) and generating electricity at voltages that are more directly useful to the customer. When this is done properly, these customers can even be off the grid entirely, as shown in figure 3–2.

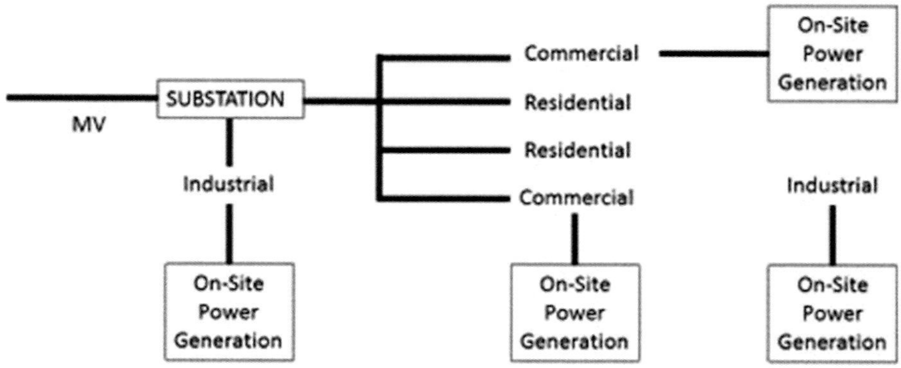

Fig. 3–2. Distributed generation

As the customer base of any given electric utility company grows, there will come a time when more power is demanded than the supply system can comfortably deliver. Most utility-scale electrical power distribution systems are interconnected in this country, so power companies can lend and borrow power among themselves to satisfy short-term power needs or surpluses. As this power shortfall situation grows more dire, considerations are given to building a new large-scale power plant somewhere—which, because of NIMBY (not in my backyard) challenges, will have to be remote from civilization. Part of this challenge is that this new power plant will require installing many miles of new very-high-voltage and high-voltage electrical transmission lines, and these can cost about $1.5 million dollars per mile to build, plus the added costs of all the real estate purchases and easements, environmental impact studies, and geologic surveys that must be done well in advance of approval of any permit applications. When you consider that utility-scale power plants are usually built many miles from any town or city, the costs of upgrading the general infrastructure can be astronomical (far beyond the costs of building the new power plant).

Many electric utility companies will rely on voluntary usage restrictions on the part of their customers, in return for a lower electricity price structure, to help their existing supply meet the ever-increasing demand, rather than taking on the challenge of raising at least a billion dollars for a new power plant. I know of one gigawatt-sized power plant in particular that has been in a holding pattern for over a decade even though the electricity it could provide to the grid is already in demand, because of the enormous costs—not just the financial sticker price but also those costs rising from the political, administrative, and regulatory requirements associated with building and operating it. Make no mistake: the appetite of venture capitalists—specifically the lack thereof—is a large part of why the electrical infrastructure of this country has not kept up with the growth of demand. Without a sufficiently low price tag, as well as a guaranteed return on investment within the requisite time frame, the appetite to fund such power plant projects by the pool of investors is very low.

Developing power plants for distributed generation, in general, and for on-site power, in particular, is manifold—and the most efficient way of generating power is by cogeneration. Generating electricity via fossil fuels, such as natural gas, creates enormous amounts of heat during the combustion of the fuel. In a stationary reciprocating, internal combustion engine—as found in most cars—the rotational movement within the engine is used to spin the shaft of an electrical generator. The heat from the engine, which is transferred to the engine's coolant, can be extracted via a heat exchanger and thus captured for other uses. Similarly, the engine's exhaust is emitted at about 1,000 degrees Fahrenheit (°F), and this heat, too, can be captured using a heat exchanger. The net result is the generation of two forms of energy from a single fuel, hence the term *cogeneration* (in this case generating electricity and heat). The entire process can be highly efficient, well over twice the efficiency of a typical utility company's power plant—which considerably reduces the costs of the energy consumed at the facility. As long as the potential customer is using electricity and they are changing the temperature of something, then they are a candidate for on-site cogeneration.

Figure 3–3 shows a basic block diagram of a cogeneration power plant. Natural gas is commonly used as the fuel supply to a reciprocating internal combustion engine that spins the shaft of an electrical generator. The engine's exhaust is put through a catalyst bed to clean the emissions before they are sent to an air-to-liquid heat exchanger, and then the cooled exhaust goes up a stack to the atmosphere. Hot engine coolant leaves the engine at about 200°F and enters that air-to-liquid heat exchanger, where thermal energy is transferred from the very hot exhaust to the relatively cooler liquid, causing the liquid's temperature to increase while lowering the temperature of the exhaust at the same time. The engine's coolant, now at about 210°F, passes through a liquid-to-liquid heat exchanger, where thermal energy is transferred to another fluid used by the customer's facility. Cold process water comes in from the facility, and warm/hot process water returns to the facility to be used in whatever processes the customer has. The engine's coolant, now at a lower temperature, is pumped to a radiator (sometimes referred to as a *dump radiator*). This radiator works very much like the radiator in your car, transferring heat from the coolant to the earth's atmosphere to reduce the temperature of the liquid to a safe level to reenter the engine. This is also a safety feature, because if the customer's process water flow is stopped for any reason, then the primary heat sink will be lost, and the radiator must be sized to be able to reject all of the engine's heat (including the exhaust heat) to protect the engine from overheating.

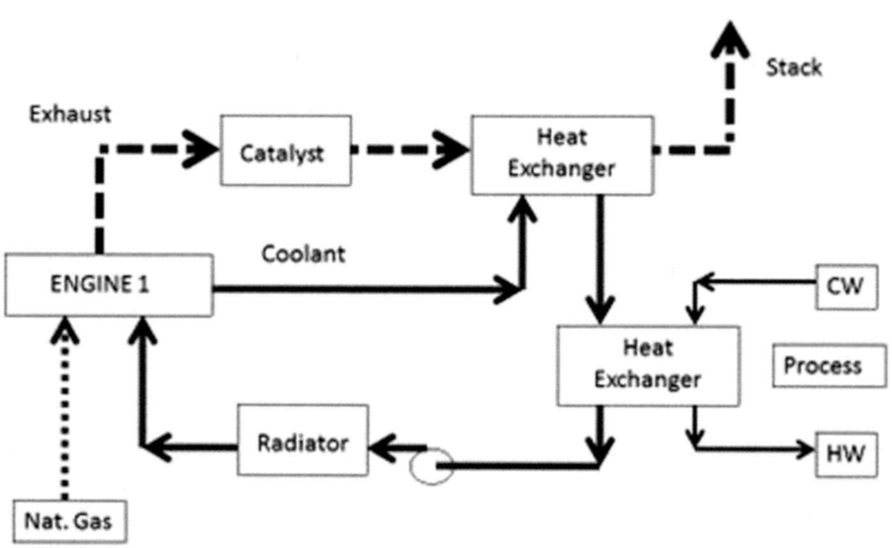

Fig. 3–3. Basic cogeneration power plant

Regardless whether the facility is a college campus, a hospital, or an industrial facility making anything from plywood to fruit juice, the same steps are used to evaluate an existing facility or to analyze a possible new one. In general, once the potential customer is identified, an analysis is performed to determine the energy needs of the customer's facility, both electrical and thermal. Several conceptual designs are created and then refined, defining the kind of power plant(s) that will meet the needs of the customer as well as the best-suited types of equipment to use (equipment selection will be covered in chap. 5). After further operational analysis, a final design is chosen, and then a financial analysis can determine if the power plant is a feasible project. If it is, then a contract is signed, the equipment is procured, the permits are obtained, and the plant is constructed and tested in the commissioning process to ensure everything works. If the given design is not financially feasible but it does meet all of the engineering needs of the potential customer, then efforts can be made to refine the solution. This could mean either finding a way to trim the budget and life-cycle costs of the chosen power plant or identifying the parts of the design that are the source of the financial problem(s) and finding alternative equipment.

Turnkey or Power Purchase Agreement

Two general paths may be followed in the world of on-site power: *turnkey* and *long-term contract.* In the turnkey situation, the power plant developer will negotiate a deal with a customer, take care of the design and engineering of the power plant, procure the equipment, and construct and perform the initial testing of the facility all before handing the keys over to the customer-owner (who handled the financing of the project). This is similar to buying a car, where the automobile manufacturer has done all of the design, engineering, construction, and testing of the vehicle before selling it to you and handing you the keys.

The alternative is for the developing company to handle everything. This includes the financing; *engineering, procurement, and construction* (EPC), testing; and all of the *operations and maintenance* (O&M) of the facility. The developing company sells the generated electricity and thermal energy directly to the customer via a long-term contract. This is typically a *power purchase agreement* (PPA), a contract that extends for as long as the projected lifetime of the equipment (12–15 years or so). This is like having your own personal taxi service, where a car, which someone else put together, is available to provide all of your transportation needs (including a driver) for an extended term (e.g., 15 years), and all you have to worry about is paying the fee (that was negotiated in the contract). Unlike Uber, though, the car and a driver are always waiting in your driveway and then again at your destination to take you to your next stop.

An option on the contract path is for the developer to sell the fully mature power plant to the customer after a year of operation following the commissioning of the plant, thus ensuring that most or all of the possible bugs in all of the various systems have been found and fixed. This option is sometimes referred to as *build, own, operate, transfer* (BOOT). There are other options for providing the EPC and O&M services that the facility needs, especially when the primary developer of the project cannot perform everything themselves. These are group efforts, with a team of several companies working together on the project. For example, there could be separate companies responsible for the following tasks:

- Design and engineering
- Finance and procurement
- Construction
- Testing and commissioning
- O&M

Joint ventures of this kind are common, since few companies can do everything by themselves without having to use a subcontractor for at least one part of the project. The benefits of joint ventures notwithstanding, this opens the door to a long-running game of allocating accountability and responsibility wherein a system or equipment failure after five or so years of power plant operations is disavowed by each party involved. While the various entities involved in the project point fingers at each other, the power plant could remain off-line for an extended period of time. Repairs can be expensive, and unless one specific entity is held responsible for the equipment failure, none of the companies involved in the project are going to want to pay for the work necessary to return the power plant to operation.

Time and Money

It takes just as much time and effort to develop a 10 kW power plant as it does a 10 MW plant. Clearly, the potential economic benefit is tied to an economy of scale, and there will be a minimum size below which the project is not economical to pursue. Every organization that develops new cogeneration power plants will need to assess their own threshold of pain for the sizing of their power plants—where it is profitable (or not) to invest the time and resources to develop a particular project. This is a financial management decision, not an engineering one, and will be a different size for each power plant developer, determined by experience or by calculation. Having made this determination allows a developer to establish an easy *go or no-go* trigger threshold, above which point the projects are worth doing and below which there may be room for some energy management consulting but not full-scale development. Once you determine the smallest-size power plant that you establish as your floor, you can use an average utility base rate (usually expressed in dollars per kilowatt-hours [$/kWh]) to determine an annual (or monthly) estimate of the cost associated with operating a facility using that amount of electrical and thermal energy as a quick metric (e.g., $360,000/year or $30,000/month). This metric comes in very handy when prospecting for new customers—few C-level executives know how much electrical and thermal energy they use on a kilowatt basis, but they do know how much money they spend on energy. If that energy expense correlates to a kilowatt amount that is above your threshold, then that prospect just became a lot more interesting.

For example, a company I once worked for established that projects under 375 kW of total electrical generation capacity were ones they could certainly do but were not worth the invested time and effort to actually do. The EPC investment in the power plant, even if they would have a long-term PPA specifying that they would provide the O&M services, would not yield a satisfactory return over the life span of the equipment. They could do everything on a turnkey basis but would not want to be the owner-operator under a long-term PPA owing to the low capital return on the investments in human power and equipment.

Once this lower limit is determined, its threshold allows for a set of easy first questions to potential customers, such as the following:

- What is your average kilowatt electrical loading on a typical day?
- What is your average annual electricity cost?
- Is the business seasonal, or does it operate year-round?

Power plant operations typically need to be at least 300 days a year for any real return on investment to be present. Food processors, in particular, can have wildly varying differences in their operational needs from season to season and even month to month. Tomato processors, for example, are ideal candidates for an on-site cogeneration power plant—but only for about 110 days a year. The rest of the year, the typical tomato processor (in California, at least) has very low electrical and thermal demands. Wineries, too, have high electrical and thermal demand during the crush but not so much during the rest of the year. For these kinds of customers, the power plant developer must either think outside the range of the standard-issue cogeneration power plant or limit opportunities to turnkey projects only.

Territorial Behavior

Identifying whether the potential customer is in a single utility's territory or whether there are options for which utility to choose is also important. Some electrical utilities are relatively easy to work with, and others are downright hostile to any prospective distributed generation entity invading their sovereign monopoly's territory. Being familiar with the utility and the potential headaches associated with working with them ahead of time will mitigate the stress of negotiating interconnection agreements and protective relay settings later.

Similarly, careful scrutiny of the potential customer is also warranted. How long have they been in business? How long have they been at this location? How secure are their finances? Are they in any kind of litigation? What is the condition of the labor/management relationship? Does the company have plans to move to another city or state any time soon? Is there reason to doubt whether the customer will still be a viable business in 12–15 years? These questions are more pertinent when the deal will be a PPA, where the *independent power producer* (IPP), whether that is the developer or another party, will be looking for reasonable assurance that they can get a return on their investment. That said, those questions are still applicable to a turnkey situation. For example, if the customer wants to move in five years to another location, making the power plant more modular in its design and construction will make it easier to take the power plant along with everything else.

Once the basic questions have been answered satisfactorily and there is interest in further analysis and evaluation of a potential customer, an easy first step is to gather energy consumption data. After executing a mutual nondisclosure agreement—to keep all the lawyers happy—I ask for their electrical load history and their natural gas data (or whatever fuel they are using to generate thermal energy), both at as high a resolution as possible, for at least one year. For the electrical history, this typically comprises 15-minute interval data, and for natural gas, typically used in on-site boilers, these data can be daily or sometimes only monthly totals.

You cannot make the data more precise than they are, although I do know one energy management consultant who tries to do so. In modeling power plants, you cannot use an *average* number as a *precise* number. That is, you cannot take a daily natural gas volume consumed, divide by 24, and use that result as an average hourly consumption (unless the facility operates at a constant thermal load 24 hours a day, and in my experience, I have never seen such a facility). There is a difference between accuracy and precision, as well as a weighted average versus a straight average.

For example, if you consume 2,400 calories of food in a day, that does not mean that you can divide by 24 and claim that you consumed 100 calories an hour, every hour of the day—that would be using an average number as a precise amount. However, if you consume three meals a day, you know the times of those meals, and they are roughly equal in duration and in their calorie amounts, you can reasonably estimate 800 calories per meal and figure out an estimated daily profile that is more *accurate* than simply 100 calories every hour—but it will not be more precise.

Natural gas usage is often given as a daily amount. When this is the case, or when an even less precise amount is given (e.g., a monthly total), correlation with the actual usage on a daily scheduled basis is required in order to refine the estimated usage profile via a weighted average. To make the daily total natural gas consumption profile more accurate, analyze exactly how and when the facility uses its thermal energy. If they operate one eight-hour shift a day and you can know (by discussion with management and the actual users of the thermal energy and by direct observation over the course of a day, a shift, or an hour) what the actual natural gas consumption is, then you will be able to more accurately estimate a higher-resolution data set. This is the equivalent of having someone follow you around all day, recording what you are doing (calories burned), what you are eating (calories ingested), and when you are doing it all (time intervals), to obtain a more accurate data set than simply stating you ate 2,400 calories that day.

Having an accurate knowledge of the relationship between the electrical and thermal components of the potential customer's facility is critical. This is important because in cogeneration, electricity and the thermal component are generated *simultaneously*. If they are not consumed at the same time, one or the other is either wasted or must be stored. It is also important to know the relative amounts of energy consumed to determine if the plant should be a top-cycle or a bottom-cycle facility. In the top-cycle plant, electrical generation is maximized, and the thermal generation is whatever the facility can make. In the bottom-cycle plant, the thermal generation is maximized, and the electrical generation is whatever the plant can make. It becomes a supply and demand issue for both commodities. That is, if you size the electrical generation to meet the customer's loads and all of the thermal output of that cogeneration facility is less than the customer's thermal demand, then an additional supply of thermal energy must be a part of the overall plan. That supplemental thermal supply can be an additional piece of equipment (e.g., another boiler), but it's usually the customer's preexisting on-site source of thermal energy.

Continuing this example, if the size of the electrical generation capacity was increased to produce more thermal energy from the plant and thereby meet their thermal load requirements, then the excess electrical energy must go somewhere. Agreements can be negotiated with the local electric utility to export the excess electrical power to the grid, but the rate the utility is willing to pay for that power varies, as well as the requirements imposed on the power plant by the utility (regarding dispatch ability and curtailment). Furthermore, any agreement by the utility to buy the excess electricity generated at an on-site power plant, usually by a contract under a PPA, will have terms, in years, that are far shorter than the lifetime of the given cogeneration power plant. From the beginning, the power plant developer must take into account the very real possibility that the given utility will not renew the PPA at some point in the future—and then what to do with the excess power? Contingency forecasting is a critical part of the planning of a new cogeneration power plant or the evaluation of an existing facility, and all reasonably possible future scenarios must be considered (for more on the sizing of equipment and the models used to get this information, see chap. 5). In some of the power plants I have examined, the owner-operators had never imagined they would be trying to operate their facility under the conditions they now found themselves in—in other words, they failed to anticipate at the start all the possible situations they could encounter.

Education

One of the challenges facing a developer of on-site power plants is education of the potential customer. With very few exceptions, the captains of industry in charge of industrial and commercial facilities who could benefit from having their own power plant simply have no idea that such a thing is possible. The vast majority of people believe that the local electric utility is the only choice of provider, as it has always been that way. Providing the service of introducing the C-level executives of industrial and commercial facilities to the possibilities of having their own power supply, for electricity and thermal energy, is a critical element of any power plant developer's business plan. This goes beyond just letting these people know that on-site power plants exist and convincing them that they need one. This education must also include what the realistic limits of these power plants are.

I have been to dozens of presentations to boards of directors where the benefits of having an on-site power plant were given. After the developer analyzes the electrical and thermal energy usage data and comes up with a preliminary design for an on-site power plant to meet that potential customer's needs, assuming the initial financial analysis is positive, the sales pitch is then made. The amount of work by the developer to get to this point is minimal. To move forward with the project, though, the developer would have to invest some money and person-hours, and before doing so, they'll want some buy-in from the potential customer. Hence, the sales/information presentation to the board of directors of the target customer. The typical businessperson will easily see the benefits of having an on-site power plant, but the presentation is almost always received with a fair amount of skepticism, as the whole idea sounds too good to be true.

There is a certain mind-set associated with the stereotypical agriculture-based businessman, sometimes referred to as the *farmer mentality*. The following are the typical characteristics associated with this mind-set:

- Taking a very conservative approach to spending money on new projects
- General distrust of financial companies
- Having a lot of questions about everything, particularly what-if questions
- Being risk averse
- Fierce skepticism about any deals that sound too good to be true

All of these are good characteristics to have as an entrepreneur and are especially prudent when applied to the fickle world of agriculture-based businesses. My point is that it should not surprise the power plant developer to be met with skepticism when making the case for having an on-site power plant to a potential customer. It's easy to anticipate the customer's arguments if you put yourself in their shoes.

As a successful business person who is not familiar with any other option than having to use to electricity provided by the local electric utility company, how would you respond if you were told by an outside vendor that they could do the following:

- Analyze your company's energy management
- Design a power plant just for your company that would meet all your current and future needs, for both electricity and thermal energy
- Procure the equipment and build the power plant at your facility, perhaps even provide the financing for the project
- Provide your site with cleaner power, in terms of the voltage and frequency, than the grid can supply
- Provide your facility with a more reliable electricity supply at a lower cost than the local electric utility charges
- Agree to a long-term contract, such that your electricity charges would not increase over time (as opposed to the way the electric utility raises their rates)
- Provide your site with the needed thermal energy, at a cost cheaper than you could make it yourself in your own boilers
- Furnish all the O&M services of the power plant, such that no one from your organization would ever have to do anything
- Provide a guarantee that your company would never pay more for their energy than the utility would charge for the same amount

This package sure sounds like the customer wins and the developer wins (otherwise they wouldn't be pursuing it), but how can everybody win? That sounds way too good to be true, and I have seen many a chief executive officer who could not accept it for that very reason. "There's something not right, here—something I'm missing" is the response from business leaders. To be sure, there are unscrupulous power plant developers who are less than 100% honest with their potential customers, and their proposed projects may very well be less than ideal solutions for the given potential customer. However, there are also plenty of other power plant developers in this country who are completely honest and aboveboard, even willing to go *open book* with the project to prove everything is legitimate.

In any case, beyond the standard development of a new cogeneration power plant, the developer must have some kind of education program. There needs to be basic and then more advanced training available to the management of the potential customers about how these kinds of power plants work. This should include not only how the plants work on a mechanical and electrical level but also, on a practical level, financial analysis, O&M concerns, and long-term impacts on both the customer's facility and the power plant of continuous operations. Safety-related issues must be discussed, covering everything from high-voltage electrical safety to spill control of on-site hazardous liquids, the use of remote emergency shutoff switches, and other potential concerns.

The developer needs to keep in mind that all of the aspects of having an on-site cogeneration power plant are new to the customer. Familiarity breeds contempt, so it is easy for the developers of power plants to not even think about matters that are true concerns to the new customer. This is not unlike the courtship phase of a long-term personal relationship—and the longer the courtship is, the happier the relationship will be over the 15–20 years of the life of the power plant. If the project is a turnkey one, then there's still going to be some time period after the sale of the power plant to the customer during which the developer is going to be providing some kind of support (even if just a warranty). The better educated the customer is at the start of the project, the fewer "But I thought" objections and questions will come up in the years after the power plant is operational.

4

Site Analysis

For Both New Projects and Existing Facilities

The analysis of the site for a potential cogeneration power plant has two main components: data analysis and physical analysis. The data analysis is very straightforward, looking at the electrical and thermal energy history (usage) and gleaning information from it, first on a macro level and then in microscopic detail. The physical analysis is that of the site itself, including the location, the condition of the buildings, and the existing infrastructure, as well as taking note of all the factors that will have an effect on the design of the power plant (e.g., dust levels and elevation).

Data acquisition

Getting your hands on the energy data for an existing facility can range from extremely easy to completely impossible, because each electrical utility will have a different way of doing things—and often there are differences even within the territory of a single electrical utility. Although many electric companies have transitioned to using *smart meters*, many have still not done so. The best you can hope for is a set of 15-minute interval data for at least one calendar year, and the worst case I've had to deal with was a monthly total of kilowatt-hours consumed with no details at all. The key is to get at least a year's worth of consumption information at as high a resolution as possible, and even where the electrical data are at 15-minute resolution, the thermal energy usage is rarely that precise. If the electrical data are not at the desired resolution, the method described in chapter 3 for finding the natural gas usage can be employed, using data logging equipment (e.g., the Fluke 435).

Data analysis

Once the historical data have been obtained from the customer (or the local electrical utility) at the highest resolution available, I start with the analysis of their electrical load profile. In the 15-minute interval data, from which information the electric company creates billing statements for their customers, the data from the electric meters could take any of several forms: it could be the peak value during that interval, the average value, or even just a random sampling during that period—essentially, some meters are "smarter" than others. Often, within a given utility's territory, a variety of smart meters are deployed so even the utility itself cannot say how that 15-minute interval data are actually being collected. Tracking down the field personnel who programmed the given smart meter for your potential customer is often difficult or impossible. Without any clarification from the local electrical utility, I treat the data as peak levels during the interval (rather than an average level).

Through review of the electrical interval data, a lot of information can be gleaned from a few minutes of work, giving a macro assessment of the energy situation. In a single column in a spreadsheet, a year's worth of 15-minute interval data will be 35,040 lines long. By plotting the data on a graph in several time segments—annual, semiannual, monthly, typical week, typical day, and so on—you can see seasonal, annual, and shift-work trends.

The annual plot often looks quite messy, although I will show later in this section that there are ways to plot the exact same data set and make the picture much clearer. For example, the annual electricity imported from the utility for a poultry processor is shown in figure 4–1, in kilowatt amounts, and there are a few things we can learn about the facility straightaway. Notice that their peak load starts off at about 800 kW, jumps to 1,200 kW, ramps up in the fall for Thanksgiving, then drops, and picks back up again for Christmas.

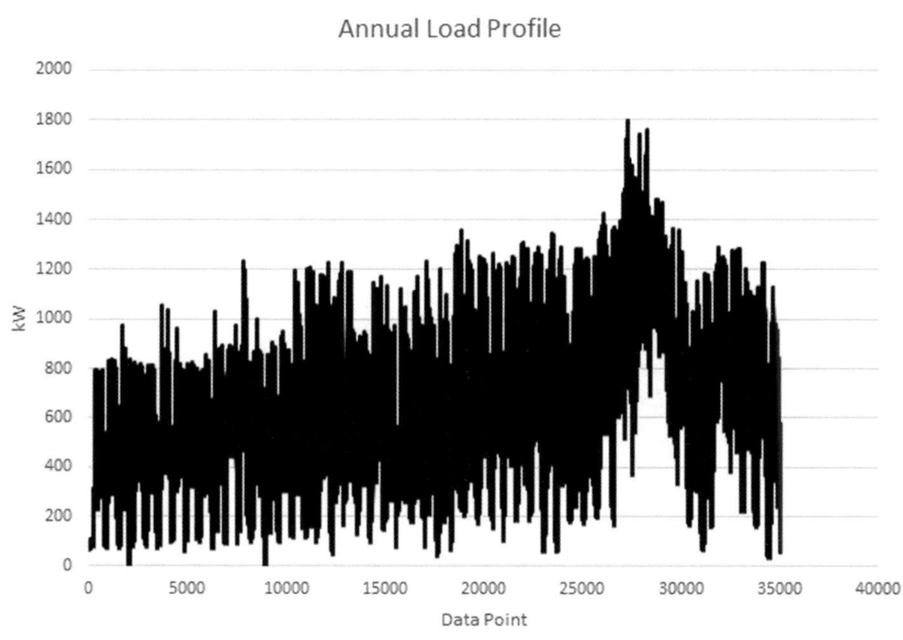

Fig. 4–1. Annual load profile, poultry processor

Now zoom in. Figure 4–2 shows the first 12,000 data points, from which a weekly pattern emerges. Zoom in further, and a definite weekly pattern becomes clear (fig. 4–3).

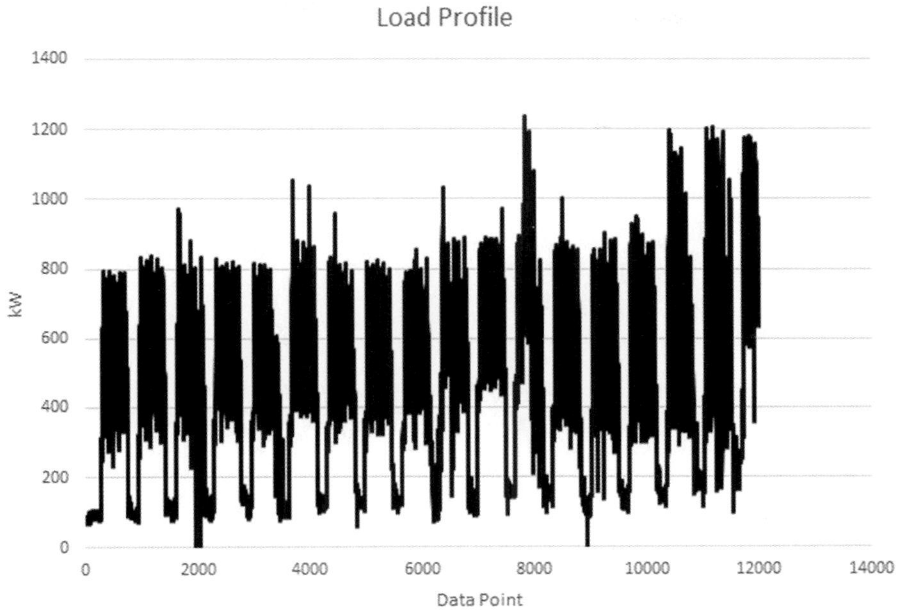

Fig. 4–2. Emerging weekly pattern

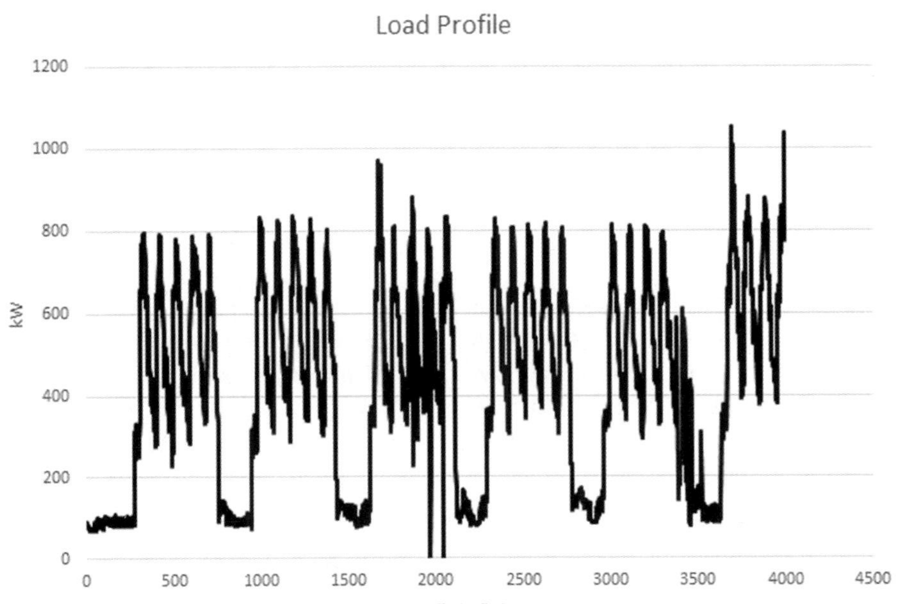

Fig. 4–3. Definite weekly pattern

Note the two points where the electricity imported from the utility falls to zero. This could be because of a failure on the part of the utility (power outage), or it could be because the facility intentionally opened the utility intertie circuit breaker to de-energize the building for some kind of electrical modification or repair. There is no way to tell which case it is from the available data.

Zoom in yet again, to a weekly profile, shown in figure 4–4, and you can see the changes in demand over the course of any given day. Cleary, the electrical load profile is not a constant, steady load. While certainly not ideal, this profile, if taken into account during the design phase, is not so erratic that a solution for generating power on-site to meet the demand is impossible.

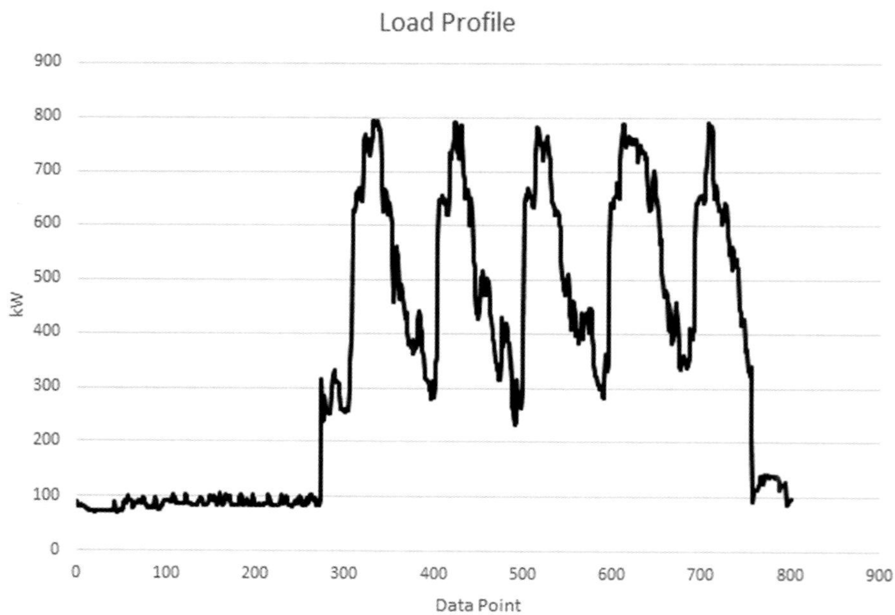

Fig. 4–4. Weekly profile

The exact same data set can be used to generate two other curves that are less messy and a lot easier to work with: the *cumulative load profile* (CLP) and the *load duration curve* (LDC). To generate these curves, copy the column of 15-minute interval data (raw data) and paste it into a new column on your spreadsheet, then sort from largest to smallest value (sorted interval data). In the next column, number the sorted data, from 1 to 35,040 (or however many you have).

The beginnings of a CLP data set are shown in table 4–1. At a glance, you can see that at 12:15 a.m. on January 1, the load is quite low (89 kW) compared to the high point (1,792 kW), regardless of when that high point is reached during the year. This difference between holiday and peak load level is important to remember when conceiving ways to generate electricity for the customer, because very high and very low load levels can be difficult to provide. In these cases, consideration should be given to shifting the various electrical sources and loads to switchgear arrangements that optimize the generation capacities and load utilization.

Table 4–1. Cumulative data set

Date and Time	Raw	Sorted	Data Point
1/1/2015, 12:15 a.m.	89	1,793	1
1/1/2015, 12:30 a.m.	82	1,779	2
1/1/2015, 12:45 a.m.	82	1,774	3
1/1/2015, 1:00 a.m.	81	1,771	4
1/1/2015, 1:15 a.m.	83	1,768	5
1/1/2015, 1:30 a.m.	82	1,767	6
1/1/2015, 1:45 a.m.	82	1,763	7

Next, choose a demand interval. I usually start with 100 kW (or 10% of peak load) for a first look before refining it; however, the smaller the interval is, the more precise your representation will be for the cumulative analysis. In a new column, establish your ranges of data and then calculate the differences between the ranges and the duration within each range, as shown in table 4–2.

Table 4–2. Developing the CLP

Demand Range (kW)	Data Point	Delta	Hours	Days	Total Days
1,700–1,800	37	37	9.25	0.4	0
1,600–1,700	76	39	9.75	0.4	1
1,500–1,600	120	44	11	0.5	1
1,400–1,500	310	190	47.5	2.0	3
1,300–1,400	935	625	156.25	6.5	10
1,200–1,300	1,936	1,001	250.25	10.4	20
1,100–1,200	3,837	1,901	475.25	19.8	40
1,000–1,100	5,879	2,042	510.5	21.3	61
900–1,000	8,535	2,656	664	27.7	89
800–900	11,430	2,895	723.75	30.2	119
700–800	14,658	3,228	807	33.6	153
600–700	17,967	3,309	827.25	34.5	187
500–600	20,835	2,868	717	29.9	217
400–500	23,893	3,058	764.5	31.9	249
300–400	27,234	3,341	835.25	34.8	284
200–300	30,577	3,343	835.75	34.8	319
100–200	34,010	3,433	858.25	35.8	354
0–100	35,040	1,030	257.5	10.7	365

The difference between the data point numbers will define each kilowatt range and, therefore, the cumulative amount of time spent within it. Essentially we are counting down from the peak imported amount all the way to zero (or whatever the lowest imported amount is). In this case, the peak was 1,793 kW, and looking down to the lowest 1,700 kW data point, we find it to be at line 37. Dividing by 4 gives us the number of hours, and dividing again by 24 gives us the number of days that the imported electricity was within that range (on a cumulative basis). Continuing down to the lowest 1,600 kW data point, at line 76, we can calculate the difference (delta) at 39 time intervals and continue with the analysis in the same manner as with the first kilowatt range. It can be helpful to start with large kilowatt ranges to get a general qualitative shape of the curves and then reperform the analysis using smaller kilowatt ranges to refine the graphs as desired; in particular, I like the 100 kW interval for facilities using 2–3 MW or less.

A couple cross-checks are warranted at this point to ensure that you have a full set of data. First, verify that the last entry in the "Data Point" column matches your last line of data (35,040 in this case). Second, verify that the last entry in the "Total Days" column is at least 365 (or however much history you're analyzing). To generate the CLP curve, simply plot the kilowatt interval (0–1,800 in this case) against the cumulative days (total days column), as shown in figure 4–5.

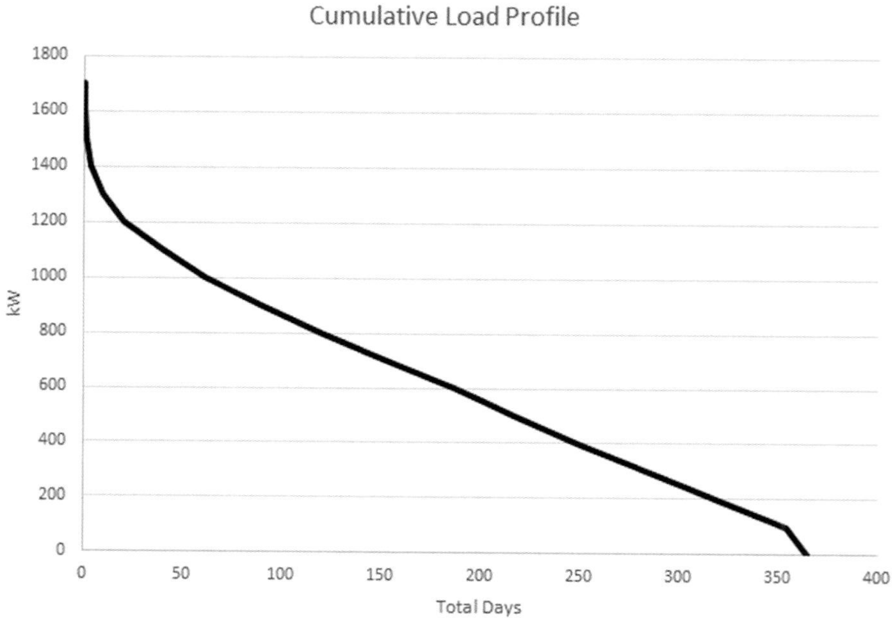

Fig. 4–5. CLP

From the 365-day CLP in figure 4–5, it can be seen that the facility spends very few days at high power levels (loads >1,400 kW) and has a wide range of loads over the course of the year. The *ideal* curve would be a straight horizontal line; hence, the higher the slope is, the less ideal the load profile is. It can also be seen that this particular facility had an electrical load of at least 400 kW for a total of 249 cumulative days that year and at least 500 kW for 217 cumulative days. The benefits to knowing this will become clearer as the next examples are described.

To generate the LDC, plot the cumulative days in each range (days column) versus the kilowatt range, as shown in figure 4–6. This shows the time, or *duration,* spent operating within each defined kilowatt interval. Here, the difference between the 400 and 500 kW points can be seen as 32 cumulative days, and again it can be seen that the facility spends very little time operating at the higher power levels.

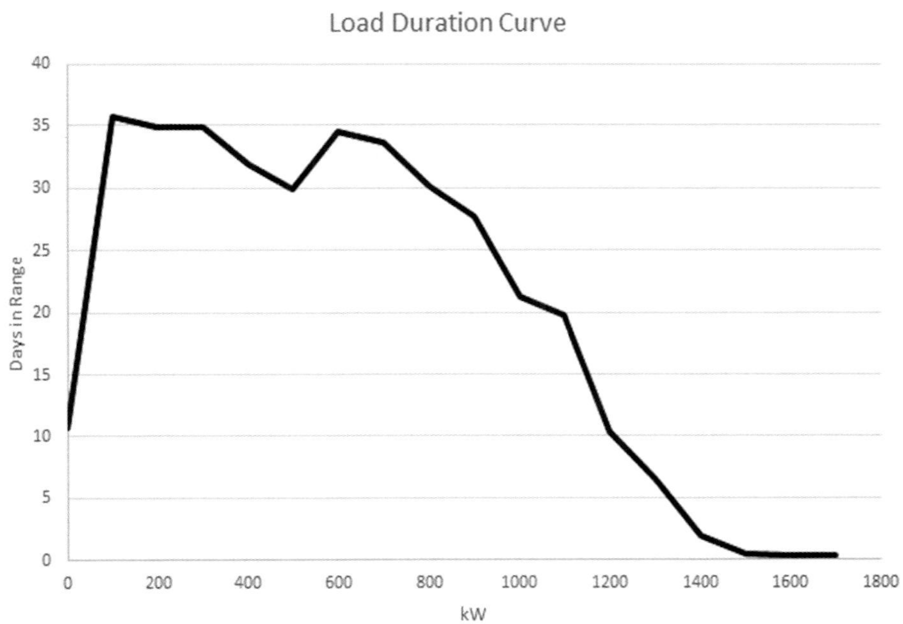

Fig. 4–6. LDC

Both the CLP curve and the LDC provide a wealth of information. This is most apparent under more ideal circumstances, as will be shown next. Figures 4–7, 4–8, and 4–9 respectively show the load profile, CLP, and LDC for a food processor. From these, it is evident that the food processor is better suited to on-site generated power than the poultry processor.

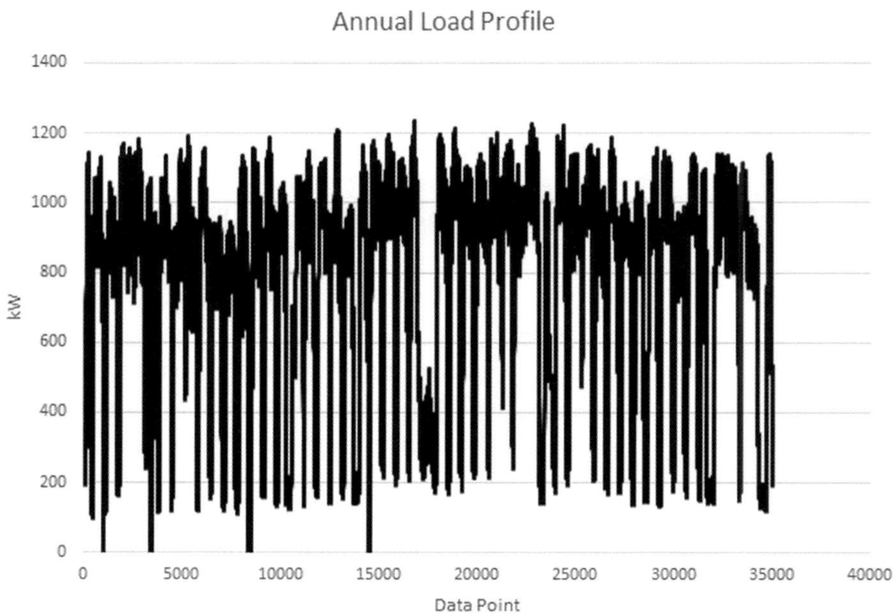

Fig. 4–7. Food processor load profile

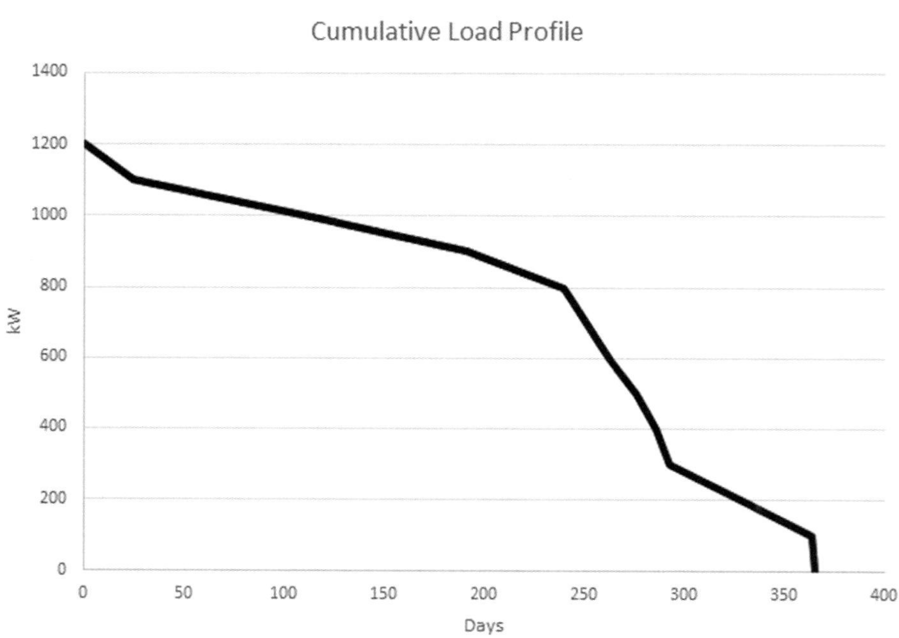

Fig. 4–8. Food processor CLP

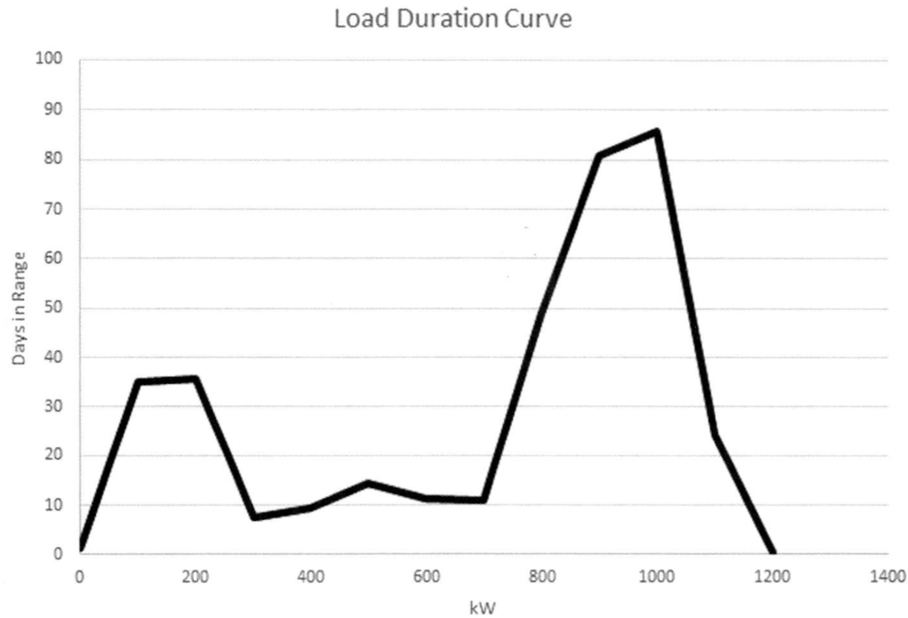

Fig. 4–9. Food processor LDC

In figure 4–8, the sections closer to a flat line indicate a more steady load, and the sections with a steep slope usually indicate a transition period between load levels. A trained eye would note that this facility operates at essentially two speeds, probably a fairly steady weekday load (900 ± 150 kW) and a lower weekend load (200 ± 100 kW). This is easier to see in the LDC, shown in figure 4–9.

The LDC shown in figure 4–9 is a good example of weekday versus weekend shift work, with most of the days of the year operating at one level of production, and less on the weekends (or second shift). The LDC can be thought of as a graph of the inverse of the slope of the CLP, such that where the CLP shows relatively flat areas, the LDC will show as peaks. Using figures 4–10 through 4–12, which are from a university campus, we can get into the details of how to use these two curves to their full advantage.

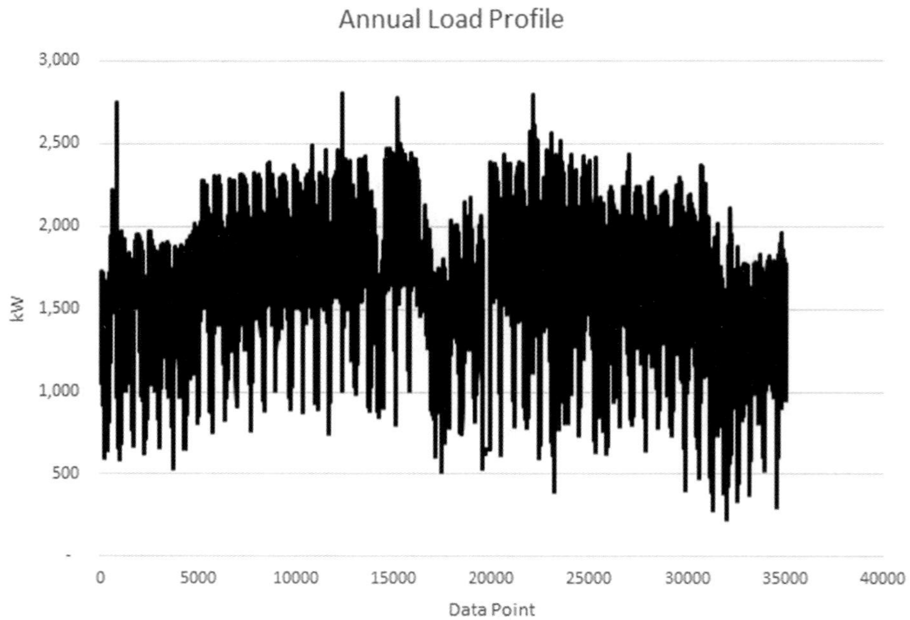

Fig. 4–10. University campus annual load profile

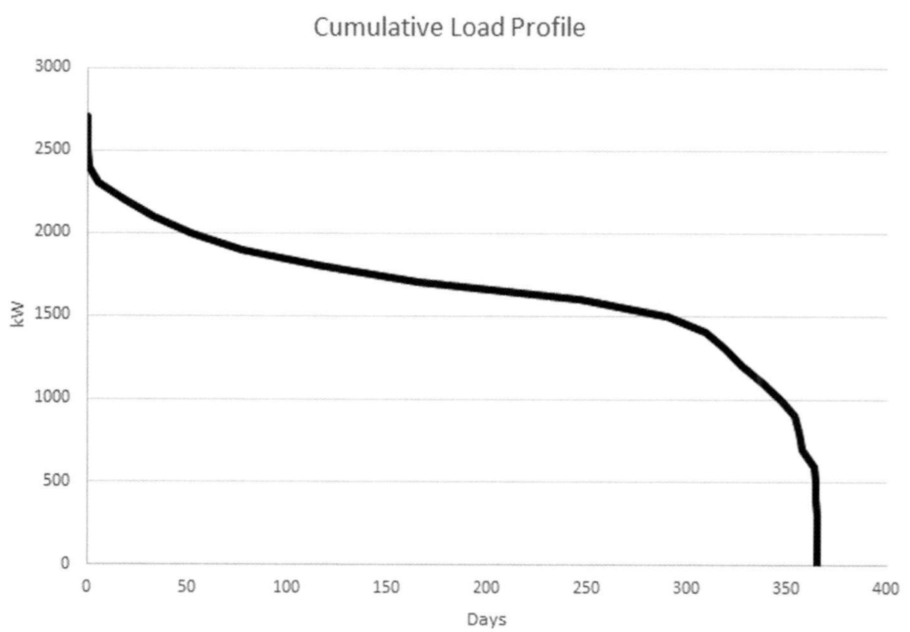

Fig. 4–11. University campus CLP

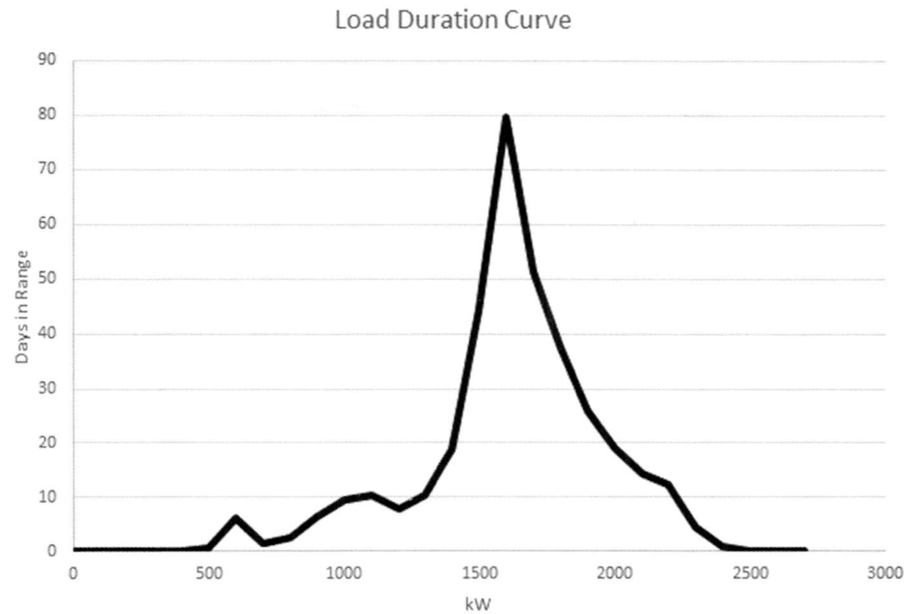

Fig. 4–12. University campus LDC

Starting with the CLP in figure 4–11, there are some things that stand out immediately, especially the two *knees* of the curve. First, note that the facility spends very little time at high power levels, above 2,300 kW (the left knee). Second, note that the load profile is almost flat from the 50-day mark to the 300-day mark—a good sign of fairly steady load. Third, note that the right knee is a gentle trend downward from the 300-day mark to the 365-day mark. Given that there are 52 weekends a year comprising 104 weekend days, this gives an indication that the facility is operating (at least on a limited basis) for some of those weekend days. Finally, when they are operating (weekday and weekend), they pretty much run at only one speed (or a narrow range of speeds)—hence the spike shape of the LDC. This kind of steady-state loading is as close to ideal as I have experienced.

While all of that is useful information, let's get to the good stuff—namely, sizing the generators to supply the load profile on a macro level. When doing so, keep in mind that the two biggest drivers of the life-cycle costs of a cogeneration power plant are the heat rate of the engines (or combustion turbines) and the cost (including periodicity) of overhauls. The heat rate of a stationary engine is simply the amount of fuel required to generate 1 kWh of electricity, similar to the gas mileage rating of a car. Just as your car gets better gas mileage on the highway than it does in the city, the engines used to drive the generators of a cogeneration power plant are more efficient at high load levels than at low load levels, usually expressed as a percentage. At operating levels below 50% of capacity, the engines become increasingly inefficient and continuous operation at very low load levels (varying between 10% and 40% depending on the manufacturer) can cause mechanical problems (more on that in chap. 5). Using the 50% threshold, we can now apply the CLP to graphically explore the ranges of generation capacity that can be employed at any given facility.

Consider the plot in figure 4–13. This is the same as figure 4–11 but with a line noting the 2,000 kW threshold and its corresponding cumulative day amount of 52. This means that if 2,000 kW of electrical generation capacity is installed, it will cover the needs of the facility for all but 51.9 cumulative days of the year. At the same time, if we allow the generator(s) to run down to 50% load, that takes us to the dashed arrow noting the 1,000 kW threshold and its corresponding cumulative day amount of 347.

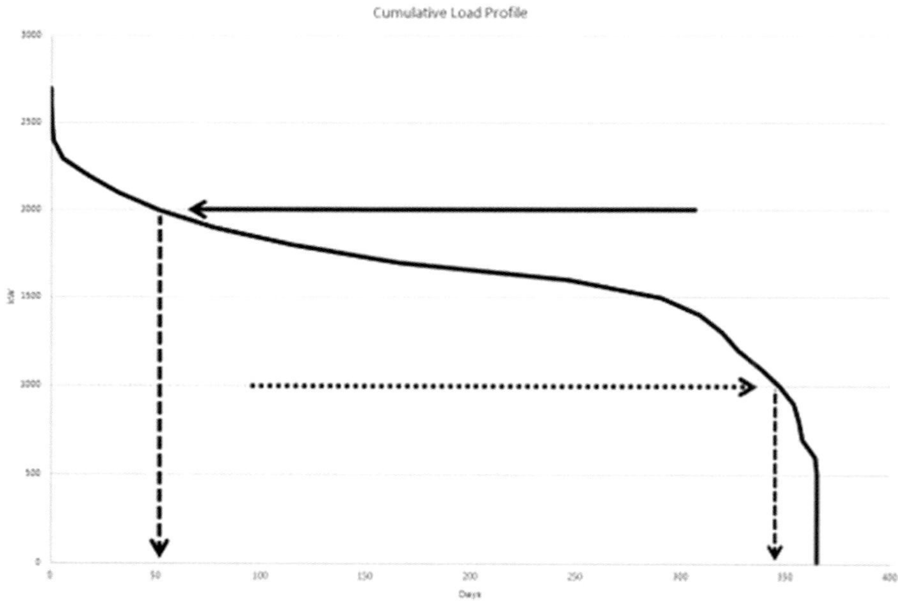

Fig. 4–13. CLP analysis

Thus, on a cumulative basis, an installed 2,000 kW generator would run fully loaded for 52 days and above 50% load for another 295 days. If the 2,000 kW of generation capacity were composed of two 1,000 kW units, then it should be clear that even more of the remaining load between cumulative day 347 and day 365 could be supplied than by using a single 2,000 kW generator (because the bottom threshold would then be down to 500 kW). Looking again at the LDC in figure 4–12 and visualizing the area under the curve between the 1,000 and 2,000 kW points, we see that it certainly is the bulk of the area, as shown in figure 4–14. If the generation capacity was two 1,000 kW units, then the area under the curve between the 500 and 1,000 kW points would also be included.

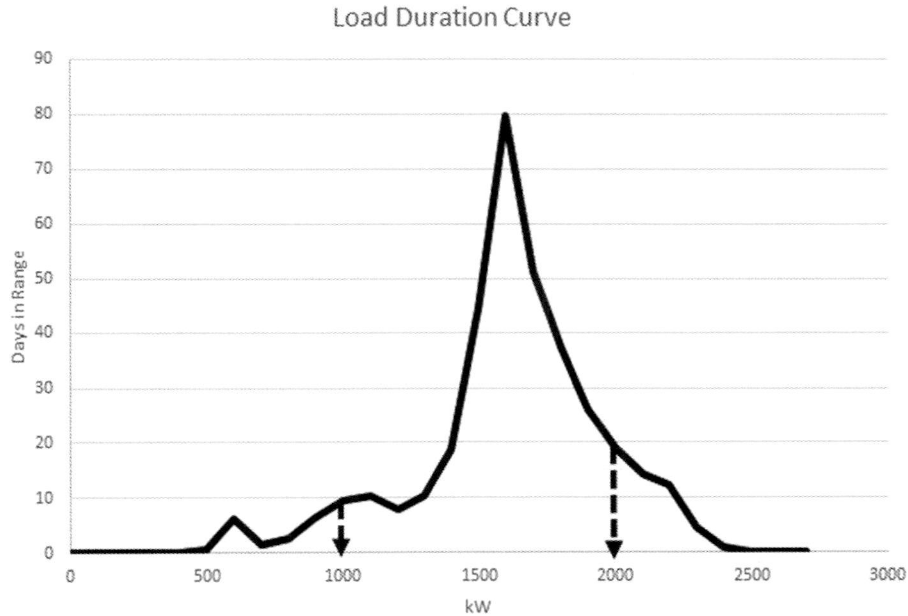

Fig. 4–14. LDC areas under the curve

Consider the area under the curve to the right of the 2,000 kW point: How can that load be supplied? Some would say to install 2,800 kW worth of generation in the first place, thereby covering all of the available load. Many would do so by using a single generator to get that generation capacity at the lowest initial cost (the economy of scale)—and that is *precisely* the kind of thinking that leads to power plants being built that can never operate at their expected levels. To see why this is so, just look back at the plot of all the 15-minute interval data and see just how much time a single 2,800 kW generator would operate at less than full load, using figure 4–15.

Note that a single 2,800 kW generator would be fully loaded only a couple of times a year and would operate above 93% load only four times a year. Also, while the generator could be run at less than 50% load, the engine that's driving that generator would be operating at a very low fuel efficiency. When you account for the increased fuel costs, increased maintenance costs, lower exhaust temperature and exhaust flow rate (and the impacts on the thermal recovery system), it becomes financially irresponsible to build such a cogeneration plant. So, while annual load profile, CLP, and LDC are simple tools to quickly estimate what a reasonable

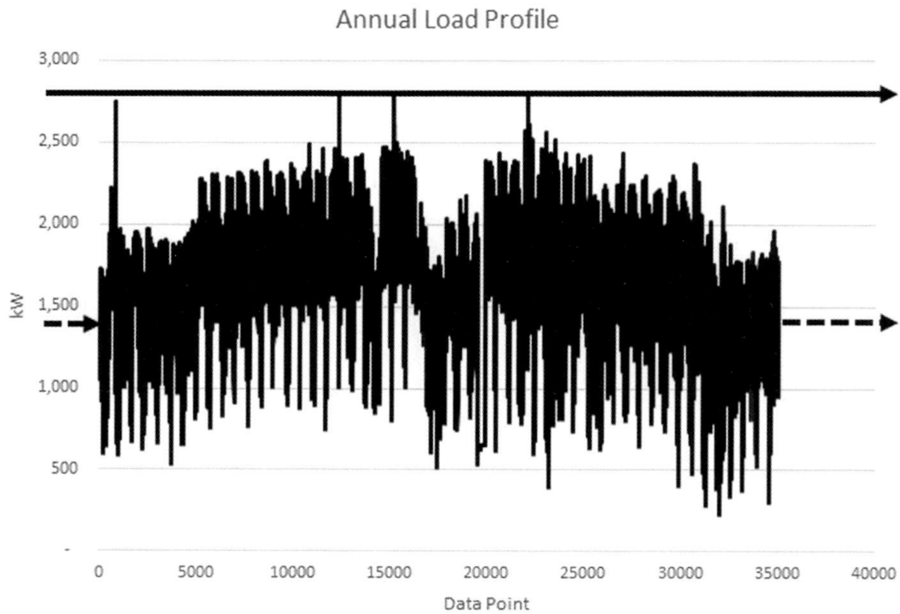

Fig. 4–15. Annual load profile under maximum generation

generation capacity is, a detailed model is necessary to refine that estimate—including the determination of how many generators to use. (This is covered in detail in chap. 6.)

Thermal Energy

In addition to the electrical load profile, the profile of thermal energy usage is also needed and then a check for thermal-electrical load coincidence. With the thermal energy consumption in units of therms of natural gas, the electrical energy consumption should also be given in amounts of energy consumed (kWh), rather than as the rate (kW) at which the electricity is supplied. For the poultry processing facility described in figures 4–1 through 4–3, figure 4–16 gives a sample month's data showing the coincidental use of natural gas (given as therms × 30, to get it on the same scale) together with the kilowatt-hours of electricity consumed.

The example in figure 4–16 shows a clear correlation in the operation of the facility. Specifically, when they are in production (weekdays), they are also using the thermal element (steam and hot water in this case). Further, during the days off (weekends), they leave the lights on (or other small electrical loads totaling about 190 kW) and do not use their boiler. Even with a high degree of coincident use of thermal and electrical energy, consideration should be given to the use of

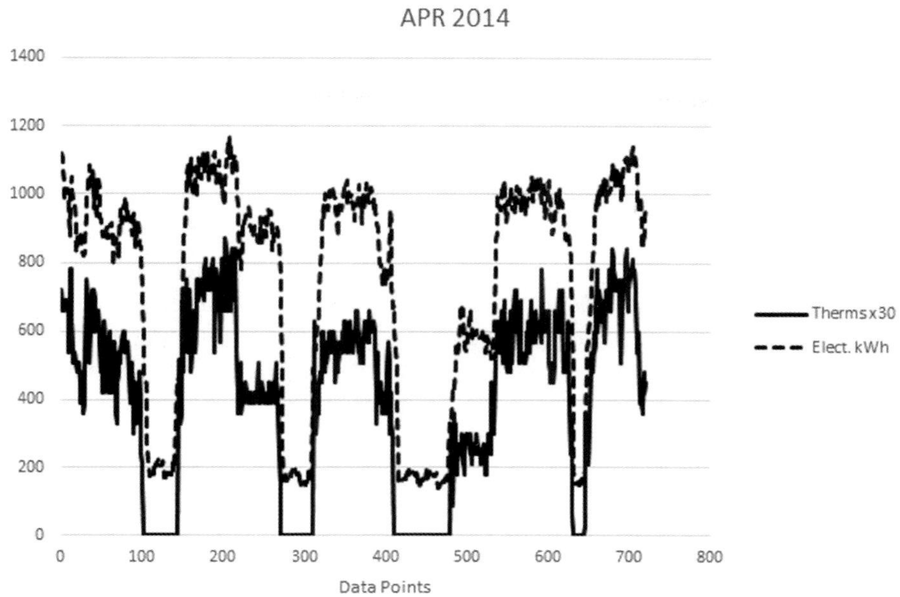

Fig. 4–16. Electrical and thermal energy coincidence

on-site storage—for either or both energies—to maximize the efficiency of the cogeneration power plant (for more detailed discussion about this, see chap. 5).

In cogeneration, we are creating the electrical and thermal components simultaneously. Thus, if they are not used at the same time, one or the other (or both) must be stored (or wasted).

Site Tour

In addition to analysis of the load profile data, a tour of the site is invaluable for several reasons. During a tour of a potential customer's facility, I take note of the following:

1. *Electrical distribution.* Note the location and condition of the motor control centers and switchgear in and around the facility, the use of variable-frequency drives, and the size of the electric motors used and nature of their loads. Are they conveyor belts, pumps, fans, air compressors, or something else?
2. *Electrical feeds.* How many electrical feeds from the utility are there? When businesses grow organically, the easiest thing for the utility to do is drop another *feed* (power line) from the local grid. I know of a dairy that has been around for a long time and now has 14 such feeds from the electric company and 14 separate electric bills. Imagine 14 individual

on-site cogeneration power plants: the costs versus the economy of scale would be ridiculously expensive, so consolidation is the way to go. That said, consolidation of all of these electrical feeds (or even a few of them) is not always possible; nevertheless, it is desirable to maximize the available load for the on-site power plant. Ultimately, if the on-site power plant is going to be intertied to the utility, then the local grid must be able to support the entire load that is to be powered by the on-site plant (to cover the eventuality of the on-site plant's going off-line). This imposes an infrastructure limit on the size of the on-site power plant. If several utility feeds are consolidated into one and intertied with the power plant and the power plant later fails (for any reason), then the single feed from the utility and its associated transformer must have the capacity to meet the needs of the customer.

3. *Utilities.* Is there grid support? What is the capacity of the existing utility power lines to the area? What is the natural gas availability (the volume and pressure), and are there any pipelines nearby? If there's already a natural gas supply to the facility, is there sufficient volumetric flow capability such that the power plant can use that same pipe? Is there any acreage for solar? Is there access to water and a wastewater treatment plant? Also, beyond just the energy and electrical distribution infrastructure, what is the layout of roads, and how can the site be accessed? Are there railroad tracks, water and sewer pipes, and existing exhaust stacks? Is there any source of biomass or biogas?
4. *Ambient conditions.* What is the annual profile of the temperature, humidity, winds (and prevailing direction), and noise levels (being considerate of nearby residences, schools, hospitals, and businesses)? Other site conditions to be considered include sunlight, rain, dust, inclement weather (hail, tornadoes, etc.), cloud cover, and elevation? All of these factors affect the design of the power plant and must be given due consideration from the beginning of the project.
5. *Thermal plant.* Where and how is thermal energy used? Is there any on-site storage? If there is a steam system, is there any condensate return? If there is, what is the temperature and volume of the condensate return?
6. *Footprint.* Where does it makes sense to site the power plant? Often, the electrical infrastructure and the thermal infrastructure of industrial facilities are not colocated. Does it make more sense to put the cogeneration power plant close to the electrical infrastructure and have a long steam or hot-water pipe to interface with the thermal infrastructure? Alternatively, do you site the plant near the thermal infrastructure and have long electrical cables to interface with the electrical infrastructure? Outdoor steam piping, especially long steam piping, often becomes much more of a condenser in the winter than the

designing engineer anticipated. Careful consideration must be given to handling the condensate in such steam piping, including recovering that hot condensate. The downside to having long electrical cabling is primarily the cost of copper, as a long cable can be a lot more expensive than a long steam pipe.

7. *All other energy sources.* Is there any on-site generated solar power (photovoltaic [PV] or concentrating solar thermal), biogas, biomass, wind, or microhydropower? How much capacity is available? How old is the equipment? Are there any plans to expand?
8. *Exports.* Is anything exported to the grid? Is there any waste stream from the processes of the facility that could be used for biomass or biogas feedstock? For example, an olive processor in California used to pay money to have the olive pits removed from the site before they realized they could burn those same pits and use the heat to make steam, saving on both on-site generated steam and not having to pay a disposal fee.
9. *Water.* Is city or well water imported to the facility? Is there any condensate return (volume and temperature) from a steam system? What are the amounts discharged to the sewer, retention ponds, or fields? Is there any water treatment system used at the facility, such as chemicals or softeners? What is the water quality, temperature, pH, and mineral content? Is fat or any other contaminant in the supply or waste streams? For example, I know of a poultry processor that uses a large tank of hot water to give the freshly euthanized birds a bath, to make it easier to remove all the feathers. Once this mission is accomplished, the water in the tank is still hot enough to contain some harvestable thermal energy but also now contains feathers and chicken fat, making it less than ideal for use in a heat exchanger.
10. *Permits.* A *permit to operate* is issued by the local air pollution control authority. A copy of all current permits to operate for all sources of emissions would give you their annual limits in terms of fuel consumed (and allowed greenhouse gas emissions) and the days per year the facility is permitted to operate. It will also identify the regulatory entity controlling the emissions from the burning of fossil fuels that has jurisdiction over that site. Air-quality boards within municipal and county jurisdictions within a given state can have widely varying requirements on greenhouse gas emissions and particulate emissions, as well as monitoring and testing requirements. In some cases, the limits are per individual sources, and in other cases, there will be a sitewide limit; there are even cases where having multiple engine-generator sets at one facility can be treated as both individual sources and cumulative sitewide emitters of pollutants that can trigger regulatory controls. This can get tricky when the facility maintains its own on-site fossil fuel burning systems, typically natural gas-fired boilers—particularly when

the on-site cogeneration power plant is owned by a different entity from the customer's facility. There have been cases in California where the regulatory agency was concerned about net greenhouse gas emissions only from a particular square mile (or even an individual acre of land), regardless of who owns the assets that are generating the emissions within that area.

Electrical switchgear

During the site tour, the general condition of the electrical switchboards and distribution centers within the facility should be noted. I have seen some equipment that looked like it was 100 years old, and I have seen facilities where, to get power from one motor control center to another, they used a collection of extension cords. Such setups can be severely limiting because the local feed from the power plant to the customer (at the point of interconnection) will have to be upgraded to the current code before the new cogeneration power plant can be installed. Often, this means that the rest of the electrical distribution system in the facility will also have to be brought up to code. This expense associated with bringing on-site power to the facility may be beyond the financial ability of either the customer or the cogeneration power plant developer. Even though it is prudent and safe to bring the entire facility up to code, it is often the case that the owners of the facility would rather continue operating as they have been, at the sacrifice of bringing in on-site cogeneration power (and all of its benefits) to the facility, rather than spending the money bringing the facility up to code. In these cases, I tell them to give me a call when their facility burns to the ground and they want to start over from scratch.

Noise

Another consideration during a site tour is noise. Noise is a frequent concern of potential customers for on-site cogeneration power even if they are located in an industrial zone, and there may be local ordinances limiting noise. As cities and towns develop economically and residential districts encroach upon industrial sectors, new noise limits may be imposed even if the cogeneration power plant was there first. For example, the orchards across the street from a fruit processing plant in California were replaced with a new housing subdivision, and when people moved into the houses they didn't mind the noise coming from the processing facility (including its cogeneration power plant) since it was well below the statutory limits. However, after the power plant was off-line for a few months during an upgrade, when it was first brought back on-line, the noise levels of the new power plant were deemed unacceptably high—even though they were lower than they had been for the original power plant. For most people, perception is reality. Knowing this, the power plant operators ran only one of the two reciprocating engines at the facility for a few weeks to satisfy the homeowners that a noise reduction protocol was in place. Once this new acoustic normal was established, they resumed operation with both of the engines, and the slight increase in noise perceived a couple hundred yards away was negligible enough to avoid any complaints. Thus, even though the power plant was ultimately able to resume full-power operations, the first three months were at lower levels than planned. This goes back to the systems engineering point about involving all of the stakeholders in the initial planning stage of the project—including making upgrades to existing facilities.

Dust

On-site dust levels are a major concern for on-site power generation—and dust can even kill a project. Even if the engine-generator set is in a container, there will be no way to keep dust completely out of the operating environment. Any of the mechanically driven pieces of equipment at the site will be subject to the dust binding to any exposed oil or other lubricant. Also, any electrical generators that are vented or otherwise open to the atmosphere will be coated internally with dust, which can lead to a reduction in the heat dissipation of the equipment and, thus, operations at higher than design temperatures. Clearly, if the dust is conductive, bad things could happen to the rotor, stator, voltage regulators, and other components (the effects of an arc from rotor to stator on a 1.1 MW generator are impressive). If periodic cleaning of the generators is not a desired part of the long-term maintenance plan, there are a few ways to keep the dust out of an exposed generator, but the only one that's any good is to install the equipment inside an environmentally controlled building. Putting equipment into containers to protect them from the environment is almost never a good idea (as discussed in chap. 5).

Keep in mind that adding generator cleaning to the list of maintenance items will result in a lot more expense beyond the costs associated with the cleaning itself. There will also be the costs from uncoupling the generator, both mechanically and electrically, from the engine and electrical switchgear. On top of that will be the cost of getting a crane and doing the lifting—twice (once on the way out to the shop performing the cleaning and again to reinsert the generator into the power plant). There's also an additional risk factor, because as a result of the movement of the generator and/or the cleaning process itself, the unit may not pass the postcleaning grounding checks. All in all, having to perform periodic cleaning of the generators will add financial burden to the budget, downtime to the maintenance schedule, and an element of risk to the generator itself.

Working with the Local Utility and Interconnection Options

The layout of electrical utilities and the distribution of power lines across the United States are summarily referred to as the grid. The grid was designed, in part, to protect the monopoly on power for the local provider(s). If you look at a map of the service areas of the various independent operating electrical utilities, you will see how they have carved out territories for themselves wherein they are the sole source of electricity for all residents and businesses. An existing customer wanting to install their own on-site power plant and thus diminish (or eliminate) the amount of power they purchase from the utility is met with resistance, as a threat to the local utility's monopoly. Local and state laws requiring the utility to allow such self-generation power plants to be built do not preclude the utility from making it difficult both administratively and financially for the local power producer to complete the project.

The inconvenient truth about new on-site power plant projects is that the developer must interact with a hostile utility that will do whatever they can to impede, discourage, or otherwise make it prohibitive for the developer to complete the project. The irony is that increasing the amount of distributed generation is exactly what will help save the large electrical utilities from their own demise owing to the impossible situation resulting from the increasing amount of solar assets supplying power to their grid.

This is best explained using an example, from California, where the grid is controlled and monitored by the California Independent System Operator (CAISO). CAISO has published a graph known as the *duck curve* (fig. 4–17) that shows the anticipated changes in the amount of electricity the state's utilities will have to provide to the grid over the course of a day. This electricity is supplied via large-scale power plants scattered around the state that use either flowing water, nuclear power-generated steam, reciprocating engines, or combustion turbines to spin their generators. From just after midnight onward, the projected increase in demand over the next several years will require an additional 1,000 MW of electricity to be generated from 2013 to 2020. The overall demand from the grid decreases until about 3:00 a.m., then slowly increases until sunrise. Once the sun rises, the amount of electricity supplied to the grid from the many solar PV farms scattered across the state starts to increase as the overall demand for power starts to level out (as household demand is reduced and business demand stabilizes). Therefore, with the increasing amount of solar-provided power available to the grid, the amount needed from the traditional power supplies decreases proportionately. The duck curve doesn't show the power generated via solar fields, only that amount required from the utilities' conventional power plants (with their spinning reserve in standby). This amount decreases over the course of the day, and as more and more solar production is expected to come on line from year to year, the amount

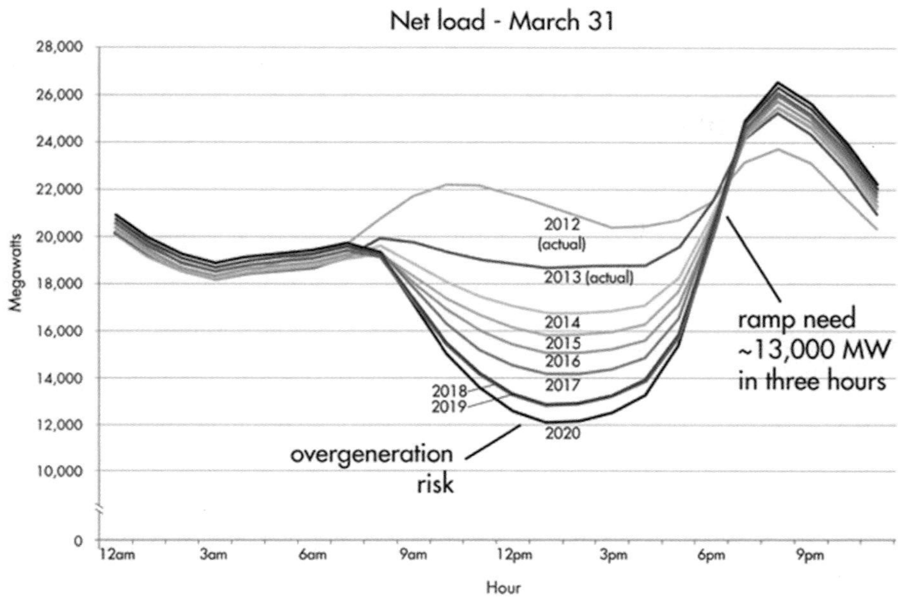

Fig. 4–17. The CAISO duck curve

(*Source:* National Renewable Energy Laboratory [NREL] Technical Report NREL/TP-6A20-65023, used with permission from CAISO)

required from the utilities' fossil fuel-burning power plants decreases further and further until about 2:00 p.m.

As the power supplied to the grid by the solar assets starts to wane in the late afternoon and the amount required to support the grid remains about the same, the amount the utilities must provide starts to increase, as can be seen from 2:00 p.m. to about 5:00 p.m. (fig. 4–17). Then, as the sun sets and the PV-supplied power begins to plummet, the real problem becomes evident. As people return home from school and work, what do they do? They turn on the lights and adjust the thermostat to turn on the heater or air conditioner; they open their refrigerators and freezers to forage for their dinner (causing the compressors to kick on); and they turn on televisions, microwave ovens, stoves, and other appliances. The net result is that the overall demand for power from the grid starts to increase rapidly from 5:00 p.m. to about 7:30 p.m., just as the abundance of solar-provided power is disappearing fast.

The utilities' *spinning reserve* power must now be brought back on line, and fast—and herein lies the problem, since there is a finite warm-up time required for equipment such as combustion turbines and steam turbines. As more and more solar PV power plants are built and supply power to the grid, it becomes increasingly harder for CAISO to manage the conventional power sources and get enough electricity supplied to the grid while the demand is spiking to avoid localized brownouts or other power interruptions. Rolling blackouts, anyone? Just look at the projected increase in demand in 2018 from 5:00 p.m. until about 8:30 p.m., going from 15,000 MW to well over 26,000 MW in just a few hours (fig. 4–17). There are many who believe the overall system will be so stressed that CAISO will simply not be able to meet that increase in demand, especially in areas of concentrated consumption such as major metropolitan areas.

This is where the traditional model of isolated, large-scale utility power plants feeding a grid of high-voltage power lines breaks down. To maintain the stability of the grid as more and more solar PV assets are brought on line year after year, public electrical utilities must embrace adding new smaller, more modular, more nimble generation components to the existing infrastructure. The most efficient way of doing this is, not surprisingly, via cogeneration power plants. Ultimately, this supports the consumer, because when an on-site cogeneration power plant is on line (and intertied with the local utility), the grid immediately surrounding the customer is made more stable for voltage and frequency variations, which benefits the local utility—the very ones fighting the installation of the distributed generation assets. Also, every 1 kW generated on-site is taking a load off the grid and is 1 kW the electric company can put to use elsewhere, which also benefits the utility.

Many electric utility companies are scrambling to add massive battery systems to their inventory of power supplies to their grid, as a substitute for building either new large-scale or distributed generation assets. These energy storage systems can be added to their existing high-voltage power lines, minimizing infrastructure costs. They can also provide support to the grid virtually instantaneously, without the warm-up time that conventional generation requires. The utility can then recharge their battery banks overnight, when the public demand is low. One problem with this approach is the question of how many eggs to put in each basket; that is, having a massive energy storage system (battery) is not a risk-free endeavor. Again, a distributed generation approach using multiple smaller energy storage facilities mitigates the risk that a single failure will have a significant impact on the overall energy supply.

Supply and Demand

The demand for power is instantaneous. You expect the light to turn on when you flip the switch, without any delay. People expect a ready supply of electrical energy the same way they expect water to come out of the tap at the kitchen sink, instantaneously and available 24/7. However, the generation of power is not instantaneous, and although some technologies can be brought on line faster than others, there will still be a delay. Many utilities are now exploring the use of energy storage systems, primarily large banks of batteries and inverters, to more quickly meet the spikes in demand resulting from interruptions in the PV-supplied power to the grid, as described above. It takes only one cloud to eclipse the sun, and the sporadic nature of the weather results in a sporadic supply to the grid from PV-supplied power sources. To ensure grid stability, a stable and reliable source of electricity must be available as a contingency at all times.

In the case of developing an on-site cogeneration power plant, the developer has two real choices: to interconnect with the local electrical utility or be *off grid*. If the cogeneration power plant will be intertied with the utility, a few options are available (or at least negotiable) most of the time (there will always be exceptions). The most common of these options is to have either a minimum import agreement, an inadvertent export agreement, or a full export agreement (that may be dispatchable). For the off-grid option, there are several options, which will be discussed in the final section of this chapter.

Coordinating with an electric utility to have an intertied generator requires that several potential points of contention be worked out, most of which are under the direct control of the utility. Make no mistake: the utility is in charge of the agreement and will dictate the terms to the developer of any on-site power plant who wishes to join the party. In addition to charging various fees—from administrative costs to ground fault and protection studies—the utility can require that certain protective relays be used in the power plant, with specific settings. In the utilities' defense, there are several matters they should be concerned about before allowing someone else's generator to be connected to the grid; therefore, studies do need to be done (e.g., ground fault), and limits need to be imposed on the allowed variation of voltage and frequency, hence the protection relay requirements.

Problems can arise, however, when those imposed relay set points are so restrictive that they end up causing *nuisance trips* of the on-site generators. Nuisance trips are cases where the protection relays cause the opening of the generator output breaker(s) in response to sensed changes in system voltage and/or frequency that are beyond the range of what is allowed but are not due to an electrical fault; rather, they are natural, short-duration electrical transients due to changes in electrical load. The issue with voltage in particular will be discussed later (in the final section of this chapter), but consider for now how the lights in your home momentarily dim when the air conditioner kicks on. This is not altogether unexpected, and it doesn't present a huge electrical problem. It would be a nuisance, though, if every time your air conditioner turned on, the main circuit breaker for your house tripped open and you lost all power.

Problems can also arise when the "protection studies" the utility requires take months to complete (with no end date set) and cost significant amounts of money up front (which is nonrefundable). This is one way utilities discourage the development of on-site power generation—the costs to the developer can be over $100,000 just to get the ball rolling with the interconnection application forms. Ultimately, if the studies performed by the utility show any problems (whether they are real or supposed concerns), then the developer is out a lot of money;

moreover, once these problems are addressed, a new set of studies will be needed, at additional expense. Even if the studies find no danger, they can take significant amounts of time to complete—leaving the developer to keep the project on hold, unable to file for permits or procure equipment until they know the project can move forward.

As an analogy, imagine you want to build a house on a plot of land you've just purchased. Before you can start construction, the city requires completion of an environmental impact study, at your expense, to be paid for up front; a down payment of $100,000 is expected along with your agreement to pay the remainder once the study is finished (though if the city doesn't use all of that initial payment, you'll get a refund). This environmental impact study will take weeks to perform, or maybe even months—there's no way to predict how long it will take—but for some extra money, you can pay, in advance, for overtime hours for the city employee conducting the study to expedite it. Also, once the city gives you the green light to begin construction, the local homeowners' association will place demands on your new house too: you'll need a security system that meets their minimum requirements, you'll need to put a fence around your yard, you can paint your house only certain preapproved colors, and so forth. Keep in mind that all of this administrative hoop-jumping occurs long before you can even finish the blueprints for your house with your architect—because they will want to add to (or modify) your design to meet their requirements—so everyone in control of the site can approve your plans before you begin.

Generic Cogeneration Power Plant

A simplified (and generic) electrical distribution diagram for a typical on-site cogeneration power plant is given in figure 4–18. Under normal circumstances, the electricity is brought in from the utility's power line, then a transformer is used to step down the voltage to 480 V alternating current (AC), and a main utility circuit breaker serves as the disconnect device (much like the main circuit breaker for your house). On the customer's side, there is a main AC bus from which the customer's various electrical loads are powered via their own circuit breaker.

When power is made on-site at 480 V AC, each generator (G1 and G2) will have its own output circuit breaker connecting the generators to the electric utility's power via the customer's distribution system. Each of these generator output breakers will be controlled by a protection relay (which comprises several relays within one unit), and there are equipment to monitor voltage, frequency, and phase rotation, among other parameters, collectively referred to as the *generator protection system* (GPS). Often, there are two GPS relays operating in parallel with each other for redundant protection. The main utility breaker will also have a similar set of protection relays, known as the *intertie protection system* (IPS). The

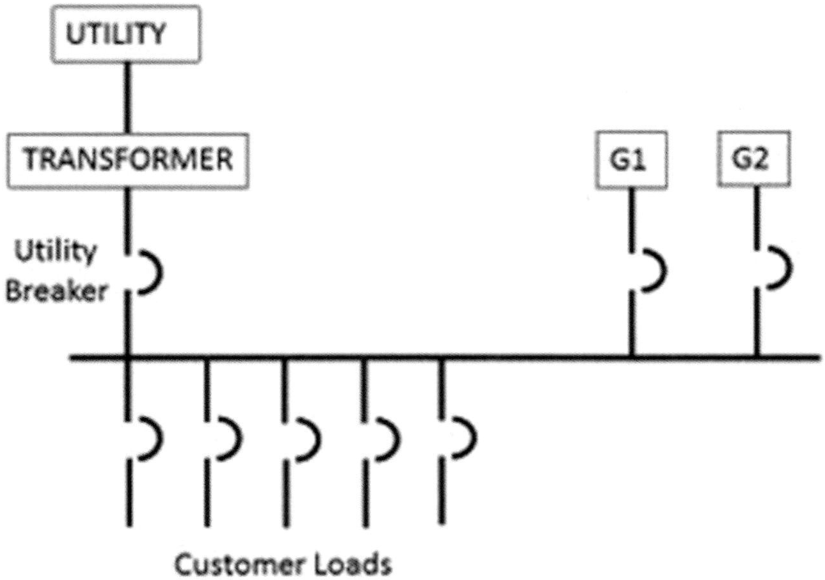

Fig. 4–18. Simplified electrical distribution

settings of these GPS and IPS relays are what the utility is interested in determining through its studies.

With the main utility breaker being shut and both on-site generators being off-line at the start, power is supplied to the facility from the utility just like the utility provides power to your home or office. The operators of the cogeneration power plant can then start G1, a reciprocating engine, or a combustion turbine (or whatever technology is being used as the prime mover) and commence generating electricity. When the generated voltage and frequency are matched to whatever the utility is providing, the generator output breaker can be closed, and on-site power can now support some of the customer's loads. As the output of G1 increases toward its maximum, the amount of power imported to the facility from the utility decreases proportionally. G2 is then brought on line in the same way, further reducing the amount of power needed from the utility. Electrical generators working in parallel are able to affect each other, and in general, the larger units will dominate the smaller units. That is, while G1 and G2 will have their own ability to control the frequency of the AC voltage they are making, it is the frequency of the AC from the utility's generators that will dictate what the frequency is for the on-site power plant.

One of the features the utility is concerned about is a ground fault. A ground fault is an inadvertent current path between an energized conductor and an electrical ground. This can be from a tree falling onto a power line, or failures of electrical cable insulation allowing contact of the wire with the metal casing of a piece of equipment, or any other circumstance that provides a short-circuit current path to ground. Once this fault occurs, there will be a lot of electrical current flowing through it, and since it's not an intentional current path, the potential for damage to equipment and/or injury to personnel is very high. The high amount of current will continue to flow until either the path is interrupted (e.g., the cable vaporizing) or the power supply is removed (disconnected).

Consider the case where a car slams into a utility pole: Either it knocks the pole down or at least one energized power line falls to the ground. "Fault 1" in figure 4–19 illustrates this example.

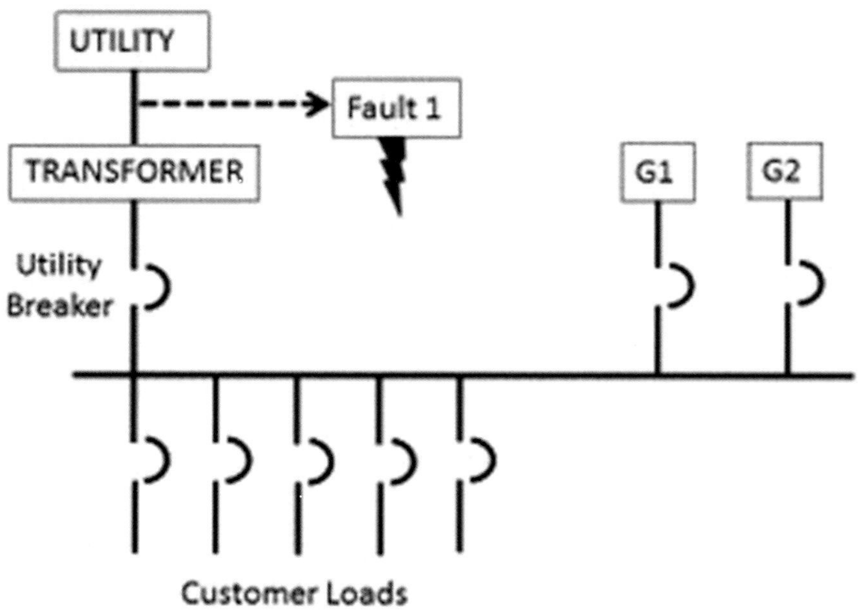

Fig. 4–19. Ground fault outside the on-site power plant

If at least one of the on-site generators is on line at the time of the accident shown in figure 4–19, power will flow to the ground fault from both the utility and, via the transformer, the customer's facility (supplied by G1 and/or G2). The intertie protection relays will sense the ground fault, as well as the increased current flowing toward the utility (instead of toward the customer), and will send a signal to trip open the customer's utility breaker. The generator protection relays should also sense the ground fault and the increased output current for the generators that are on line and will cause the generator output breakers to trip open if the GPS settings are exceeded. On the utility side, the local substation providing power to the power line will also sense the ground fault and increased current flow and may or may not automatically trip open the supply breakers. Sometimes the utility must send someone out to visit the substation and de-energize that local part of the grid manually. The reason the utility is concerned about the intertie protection relays (in selection of both the equipment to install and the settings to use) is that the utility needs to be comfortable with how large of a ground fault will be sufficient to cause the customer's utility breaker to trip open—and thus prevent the on-site generators from feeding the fault current by back-feeding the transformer.

What if the ground fault occurs on the customer's side? This scenario is illustrated in figure 4–20. Again, the concern of the utility with the selection and settings of the IPS relays is to ensure that the customer's utility breaker will be tripped open to prevent the utility's power from feeding the fault current and preventing the ground-fault condition from being felt by everyone else on the local grid. The GPS relay should also cause the generator output breakers to trip open, thus de-energizing the fault while protecting the generators from the voltage, current, and frequency transient.

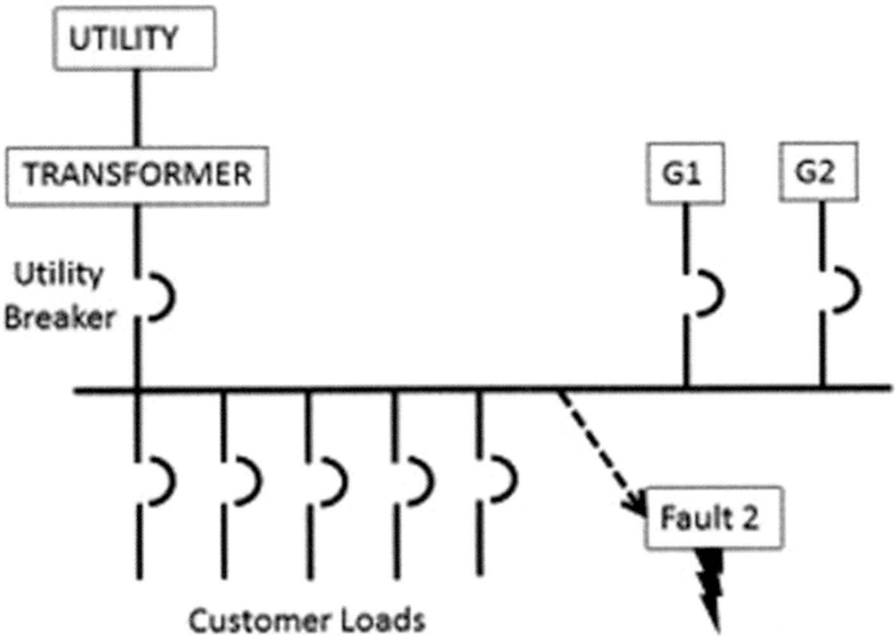

Fig. 4–20. Ground fault within the on-site power plant

There will be a coordination study as well, where the utility will examine how the circuit breakers and the protection relays coordinate their actions to maximize grid stability. In some cases, conditions sensed by the IPS relays may cause the generator output breakers, but not the intertie circuit breaker, to be tripped open. In other words, if the IPS relays sense an electrical fault on the grid, the local utility will always suspect that the problem is being caused by the on-site generators and not the utility's power plants. The first power sources to be removed from the equation at that point are the ones at the customer's facility, and then, if the problem persists, the intertie breaker will be opened. Naturally, if the problem was on the utility's side of things from the beginning of the fault, the customer's generators did not necessarily need to be shut down. Especially if opening the intertie breaker would work to protect the customer's electrical distribution system from the fault, the on-site power plant then could remain operating in island mode, allowing the customer to continue doing whatever it is they do.

Minimum Import Agreement

One way a utility can protect its own interests, as well as the grid under their control, is to require a facility that installs the ability to generate power on-site to take a minimum amount of power from the grid. That is, the on-site power generation system cannot supply 100% of the customer's electrical demand, even if they have the capacity to do so. A typical requirement is that a minimum of 10% of the customer's electrical demand at all times must be supplied to the customer by the utility, and the rest can be supplied by the on-site generators. This is known as a *minimum import agreement.*

This agreement does more than just provide the utility with a small source of continued revenue. How does the electric utility protect itself and the local power grid from rogue power inputs by distributed cogeneration power plants? One way is to establish minimum supplies by the grid to the customers, such that, if the imported power drops below some target threshold, the IPS relay will send a signal that causes at least one of the on-site generator's output breakers to open. This immediately reduces the amount of on-site generated power, which causes the imported power from the utility to instantly increase by the same amount. This kind of built-in tripwire prevents the customer from ever being able to generate 100% of their electrical demand.

The AC power that's on the grid is supplied by many different sources, and the demand for power is highly variable at the local levels (less so at the county and state levels). The *independent system operator* (ISO) is the entity that runs the grid, controlling the supplies to the grid to match the demand. The control centers of these ISO facilities look like something out of the Space Age—lots of large video screens in a mission control room, with numerous personnel monitoring every occurrence in real time and anticipating what will be needed next. They monitor what's happening not only in their part of the nation's grid but also in every other part of the system they are connected to. In this way, ISOs can borrow or lend electrical power to and from their neighbors.

Distributed energy resources, such as an on-site power plant that can export power to the grid, have the potential to disrupt the whole system. If the on-site power plant is small, perturbations will be limited to the local area around the site, but there will be a size limit above which the ISO will want some degree of control over the on-site power plant just as they have with the utility's power plants. If the local electrical utility is also acting as the ISO for the region, they can remove the variable of the on-site generators' ability to supply the grid by requiring the customer to have some minimum power import. In these cases, as part of allowing the cogeneration power plant developer to connect their on-site plant to the grid, the utility will dictate that the IPS relays are set such that, any time the amount of electricity flowing into the customer's facility drops below a certain point, a signal will be sent to trip open their generator output breakers.

Inadvertent Export Agreement

The *inadvertent export agreement* is very similar to the minimum import agreement, except that the required minimum import of power is reduced to zero. This way, the on-site power plant can provide 100% of the customer's electrical demand while remaining in parallel with the local electrical utility. That is, the customer's utility breaker remains shut, even though power may not be flowing into the facility from the utility.

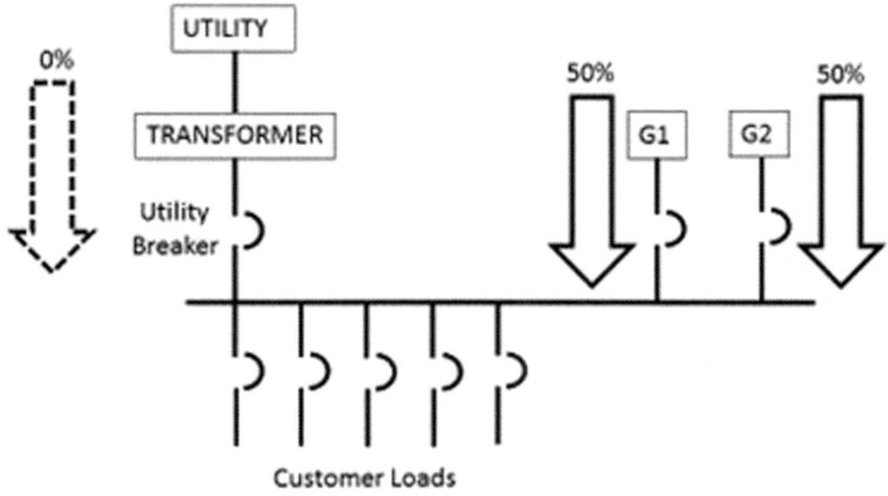

Fig. 4–21. Self-generation of 100% demand

For example, in figure 4–21, both generators are on line, each supplying 50% of the customer's total electrical load with the utility providing zero power. It's lunchtime, so all of the customer's employees stop their work and switch off their equipment (motors, compressors, conveyer belts, etc.) all at the same time; or, if just one very large electrical load is turned off, that would cause the same sequence of events to happen. Now the overall demand for power is less than it was just an instant ago, but the generators are still supplying the same amount of power as they were doing before—that is, there's more supply than demand. The drop in demand is instantaneous, but the drop in generated electricity is not. If the generators are operating in a *load-following* mode (more on that in chap. 7), the engine-generator control system will recognize the transient and throttle down on the fuel to the prime mover to reduce the output of the generators to meet the new, lower demand. Until the generators have reduced their output to meet the new demand, the extra power they were generating has to go somewhere, and the grid becomes like an electricity sponge and absorbs the excess power—hence the term *inadvertent export,* because power is exported onto the grid by the on-site power plant but only on a transitory basis. As part of this agreement, essentially as minor compensation for the reduction in revenue to the local electric utility, the on-site power plant is not usually compensated for the power that it exports onto the grid.

Full Export, Net Metering, and Dispatch

When sizing a new cogeneration power plant, there are two approaches you can take: *top cycle* and *bottom cycle.* In the top-cycle approach, the power plant is sized for the electrical load of the customer, and the waste heat is harvested to make as much thermal energy as possible. In the bottom-cycle approach, the power plant is sized to generate as much thermal energy as the customer needs, and the electricity is the by-product. In the case of a bottom-cycle plant, given the nature of the efficiencies of both the electrical generation and thermal energy-harvesting processes, if the thermal demands of the customer are higher than the electrical demands, then there will be excess electrical power once the thermal demands are met.

There are many different industries where the demand for thermal energy is higher than the demand for electrical energy, such as pulp and paper mills. When these facilities employ the benefits of on-site power generation, they will typically have a long-term export contract with the local electrical utility and sell the excess power to the grid. The problem comes when these long-term contracts are up for renewal, usually 5, 10, or even 15 years after the on-site power plant was built. The utility is under no obligation to continue to buy the excess electricity or even allow it to be put onto the grid, but the facility will still need its thermal energy to do whatever it is they do—and the only way to get that thermal energy is to also make electricity. The capital expense was made long ago, and the facility needs to continue operating—but what to do with the excess electrical power? Of course, this puts all the leverage into the hands of the local utility, and they can then negotiate as low a price as possible for the power that the facility needs to sell to the grid. This happened recently in Southern California to two large facilities; not having any other way of disposing the excess electricity, the companies involved were forced to agree to the terms issued by the utility.

Another consequence of exporting power to the grid as part of normal operations of the on-site power plant is that the power will be dispatchable. That is, the local electric utility and/or the local ISO will reserve the right to control the amount of power added to the grid. This is usually done via a direct line of communication that is established between the on-site power plant and the controlling authority, whereby the ISO can demand that the amount of power exported to the grid be curtailed (or even reduced to zero) for as long as the ISO desires, whenever the ISO desires. In other words, the power coming onto the grid from all of the various sources are under the dispatch control of a single authority. This goes for all sources supplying power to the grid, including all PV farms—anything putting power on the grid can be told not to. The flip side is that when power is needed, the ISO can request that the sources increase their production; however, the solar farms are not usually beholden to that since they cannot control the weather and make the sun shine whenever they want.

Energy Storage

One way to increase the stability and reliability of an on-site cogeneration power plant is to include in the design an energy storage element (discussed in detail in chap. 5). For example, consider the electrical distribution system shown in figure 4–22.

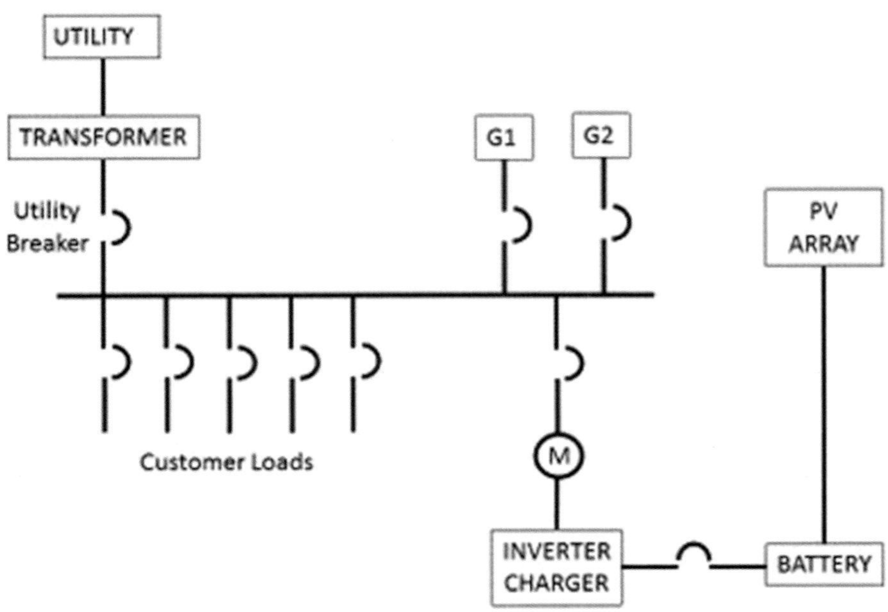

Fig. 4–22. Energy storage concept for an on-site cogeneration power plant

Added to the distribution system presented earlier are a PV panel array feeding direct current (DC) power to a battery that, via an isolation breaker, feeds DC power to the *inverter-charger*. The inverter-charger is a hybrid device that can be used to generate AC power from a DC source, or switch directions and make DC power from an AC source. Thus, during the day, the PV array can be used as an additional source of energy, and the inverter-charger can supply AC power to the customer as if it were another generator; late at night, when the customer's loads are lower and/or the opportunity presents itself, the on-site generators can increase their output and supply AC power to the inverter-charger to make DC power to recharge the battery as necessary. The PV array can also recharge the battery after the sun rises, but only if the weather allows it and the customer does not need PV-supplied AC power.

Thanks to certain state laws and other imposed incentives, some electric utilities are required to buy power generated by PV facilities supplying the grid, size notwithstanding. The key is that the rate the utilities pay for power generated from renewable sources is higher than what they will pay for conventionally generated electricity, as dictated by state law. From the utility's perspective, how are they to tell how much power is coming from the PV array and how much is coming from G1 and G2? The answer lies in a few extra power meters that must be installed into the electrical distribution system, denoted in figure 4–22 by the "M" inside the circle between the inverter-charger and the main AC bus. This way, all of the AC power generated by the inverter can be accounted for and differentiated from the AC power from G1 and G2—that is, as the net power from renewable sources. This is commonly known as *net metering*, although electrical distribution systems and metering arrangements can be quite complex.

Another advantage to having a battery and charging system is that during times when the facility is operating at less than full load, the generators can be run at higher load with the extra energy going to the storage system. Consider the 10 days of electrical load shown in figure 4–23. As before, if 2,000 kW of electrical generation capacity is installed at the facility, there are times during the five workdays when the total load exceeds 2,000 kW, and it is during these times that the battery and PV system can be used as a power source to provide the missing power via the inverter (the circled area labeled "Battery discharge" in fig. 4–23). Furthermore, when the customer's demand is low, especially below the 50% threshold, the generators can be run at a higher output and use the inverter-charger to insert power back into the battery—thus both recharging the battery and enabling the generators to operate at a higher-than-otherwise load level, improving efficiency and the heat rate (the circled area labeled "Battery charge" in fig. 4–23).

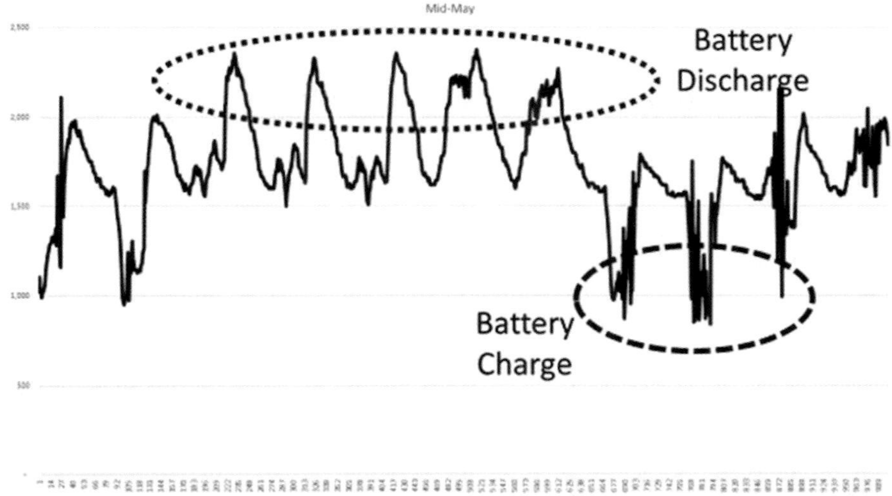

Fig. 4–23. Application of energy storage in an on-site cogeneration power plant

For minimization of the amount of electricity imported from the utility when the total load is greater than the 2,000 kW generating capacity, the inverter-charger must be large enough to provide as much of the difference as possible. The battery, too, must be sized to handle as much area under the load curve (between the load level and above the 2,000 kW line) in terms of ampere-hours (Ah) as it can.

There is a financial balance between having sufficient battery capacity to minimize the power required to be imported from the utility and the cost of that imported power. If money and the physical size of the battery are not concerns, then there's no problem. However, in real life, money is always a concern, and often the size of the footprint of the power plant (including the energy storage components) is also a constraint. The trade-off between wanting a large battery system and wanting to minimize the cost of the power plant is easily incorporated into the performance and financial models, so the operational and financial impacts can be analyzed during the design phase.

Island Mode, Intertied and Island Mode Capable, or Off Grid Entirely

What is the risk to the customer if power is interrupted? What if power is lost completely? What about if the loss is days or even weeks in duration? If the customer's electrical loads can be split into *vital* and *nonvital* designations, then a battery-inverter system can be used to supply the vital electrical loads to mitigate the customer's downtime until more generation capacity can be restored.

If intertied with the electrical utility and the utility fails, then the nonvital electrical loads can be shed to bring the customer's electrical load to within the capacity of the remaining on-site electrical generation operating in island mode.

Island mode/off grid

Being off the grid used to mean something like being at a log cabin deep in the woods, with no electricity or running water. Now, there is renewed interest in having a commercial facility that has its own power plant, such that the facility isn't reliant on the public electrical utility at all, or at least has the ability to disconnect from the grid. In the wake of recent superstorms, such as Hurricane Sandy, which devastated the northeastern United States in 2012, critical infrastructure entities such as hospitals and military installations are looking at options to be self-sustaining when conventional utilities become unavailable. Although this is certainly possible, it is also far more complex a problem than most people believe. The following are the principal problems associated with an off-grid power system:

- Transients and starting surges
- Load versus 100% generation capacity
- Reliability tree and using a modular approach
- Hybrid systems with energy storage
- Using inverters and inverter output ramp limits
- Resynching to the utility
- Fuel security

Transients and starting surges

This is the problem that plagues most power plants trying to operate in island mode, and although not impossible to overcome, it can be expensive to do so. If you've ever been in a house where the lights dimmed when the air conditioner kicked on or the lights in the bathroom dimmed when a hair dryer was started, then you've experienced the problem on a small scale.

Most of the electrical loads in industrial facilities are motors of some kind and therefore are inductive in nature. When electrical power is first applied to a motor, the shaft of that motor is not rotating. A three-phase induction motor is constructed almost exactly the same way a three-phase electrical generator is. The main difference is that instead of the rotor's being turned by some prime mover in order to generate electricity in the stator, in this case things happen the other way around: Electricity is applied to the stator, which causes the rotor to spin, thus allowing the motor to do something. Whether it's a pump, a compressor, a fan or causing a conveyer belt to move, it is still just rotational movement that is put to good use.

This is not a book about motors or motor theory, but there is a significant operational issue that needs to be addressed. The interaction between the rotor and stator is due to a rotating magnetic field, and that magnetic field is produced by the flow of electrical current through the windings of the motor. There's a funny thing about windings: when voltage is first applied to a winding, current does not immediately begin to flow. First, a magnetic field is created, and as the magnetic field gains strength, more and more current can flow through the winding. The initial inrush of current is much higher than the nominal running current of the motor, to build these magnetic fields. Once the magnetic fields are fully established, the electrical current flowing through the winding is back down to the nominal amount. This means that the motor requires a surge of current at the start, and it is this surge that is the problem. For a typical motor, the starting current surge can be 5–8 times the amount of the nominal running current; for some newer, hyperefficient motors, starting surges can be up to 20 times the nominal amount. In the planning of a new industrial facility that will have an on-site cogeneration power plant, particular attention must be paid to the size and type of motors that are included.

Ohm's law states that the voltage (V) across any given resistance (R) is equal to the amount of current (I) flowing through the resistance times the amount of resistance, or $V = IR$. This is the amount of voltage *dropped* by that given resistance. Figure 4–24 gives an extremely oversimplified example using a motor. The *motor voltage* will be whatever the *supply voltage* is minus however much is dropped by the *series resistance*. When the motor is first started, the inrush of current to initially build the magnetic fields causes an increase in the voltage dropped by the system components upstream (the series resistance). The source voltage remains the same, so the motor voltage is reduced, which is why voltage-dependent objects like lights will have a reduction in their output until the transient is over (hence the momentary dimming when the air conditioner kicks on).

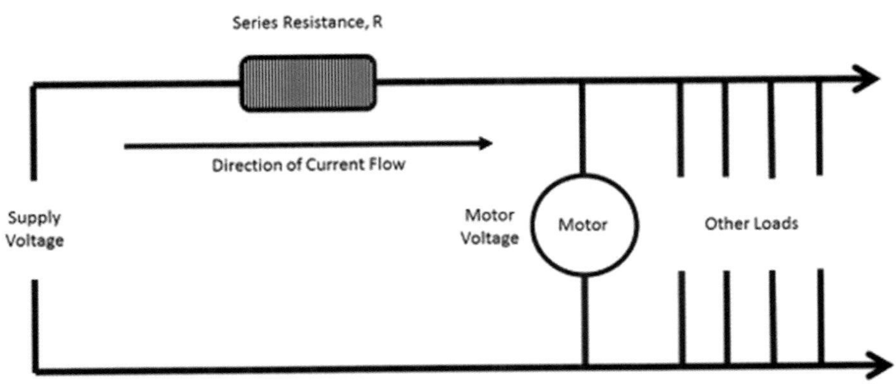

Fig. 4–24. Basic motor operation

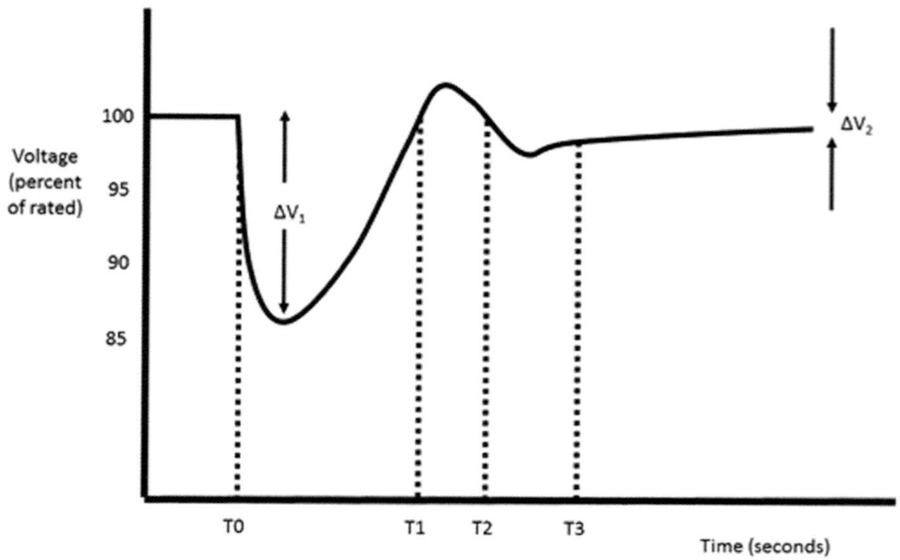

Fig. 4–25. Voltage after a sudden increase in load

This electrical transient, in terms of voltage, is illustrated by figure 4–25. Understanding this transient is the key to having a successful off-grid power plant. Knowing the magnitude of the initial voltage dip, as well as the time to voltage recovery, is vital, and it is also important to know the magnitude of the subsequent voltage overshoot and final system voltage. Everything sensitive to voltage that is being powered by the site's generators will experience this phenomenon, and some items do not react well to these changes. For example, in the food processing industry, the equipment that handles the filling, bottling, and packaging of material is typically computer or PLC (programmable logic controller) driven. These control systems can be tripped off-line by a transient voltage dip, even if it lasts only fractions of a second, bringing to a stop the customer's processes. Such stoppages, of course, are contrary to the purpose of having an on-site power plant capable of operating in island mode. In many cases, any interruption in the filling process, however short, requires a period of cleaning and resterilizing the equipment to ensure sanitary conditions are met. I once analyzed a maker of baby food, whose filling and packaging lines could not experience even a slowdown in fill rate without triggering such a cleanliness requirement. In such cases, the requirements of the on-site power plant to provide continuous electrical power at near-constant specifications are critical.

Continuity of power

The on-site power plant must be able to provide the starting current surges of the equipment used in the customer's loads without experiencing significant voltage dips, to maintain any real continuity of power while in island mode. This is where the behavior of the motors, compressors, pumps, and other components that make up the load profile must be well understood—and, better yet, documented.

Portable electronic equipment can be used to monitor the operating characteristics of individual pieces of equipment, or even entire *motor control centers*. These devices can record all of the parameters of interest at very high resolution for days or weeks at a time, providing all the real-time data a power plant designer could want. Nevertheless, this only applies to existing facilities. For the planning of a new site, using the worst-case, locked-rotor currents for the equipment listed in the plans is about the best a designer can do.

Most manufacturers of generators, when pressed, will surrender the information on their equipment detailing the response of their control systems and voltage and current regulators (which are often proprietary) to step changes in electrical loading. Using the rules of thumb that are popular in electrical engineering circles will get you close to the margins you will want, but if you use these approximations, it is wise to include a contingency factor in your calculations. In general, the smaller the kilowatt demand of the given load is, as a percentage of the kilowatt amount available from the generators supplying the electricity, the better. What happens is often one of two things, to achieve the margin of power generation necessary to ensure the largest electrical load(s) can be started without a problem while the power plant is operating in island mode: One, excess generation capacity is used, resulting in the engines driving the generators being operated at levels below full load. As described earlier, this results in lower fuel economy, lower thermal recovery, and other sacrifices. Or two, a fast-response system of backup power, such as a large-scale *uninterruptible power supply* (UPS) system, is used. This brings a lot of extra equipment and an energy storage system to the party and all the expenses associated with that, as well as requiring a larger footprint for the power plant in the first place. Again, the extra cost of having such UPS systems must be weighed against the benefits of not having to worry so much about the voltage transients on the main AC bus.

Timing and balance

The electrical transient due to the starting of a large load is not a long-duration event. The condition can be either prevented in the first place or compensated for afterward to keep the initial voltage dip from getting too low.

One method of doing this is to use a series of inverters that are supplied by banks of batteries. Keeping the inverters on line, tied to the main AC bus and in parallel with the regular generation assets that are supplying power to the bus, but leaving the inverters at very low load levels will enable them to act as a reserve. In the event of a voltage dip, the inverter control system will sense that dip and immediately increase the output of the inverter to compensate. Each inverter, though, will have a limit on the amount of power it can provide in an instant, as well as a finite rate of change of its output—just as your car cannot instantly go from 0 to 100 miles per hour (mph). One way around these limitations is to use multiple inverters, all tied to the main AC bus. To spread the reliability risk, use a separate, smaller battery bank for each inverter rather than a single, large DC source of power. This is shown in figure 4–26.

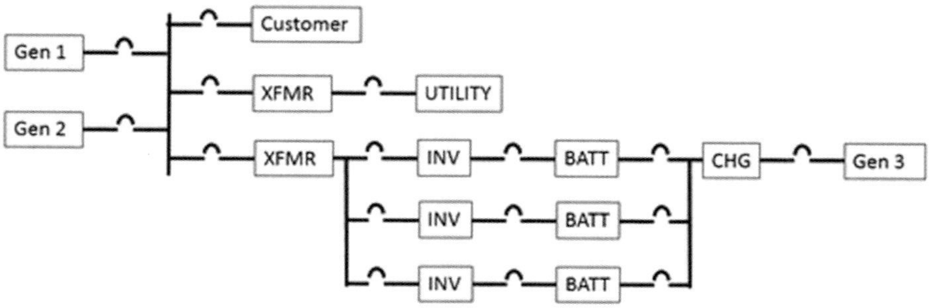

Fig. 4–26. Inverter-based transient mitigation concept

In this case, a third engine-generator set ("Gen. 3" in fig. 4–26) is used along with a battery charger. This additional set can be much smaller than the two units used to provide prime power to the main AC bus. As such, when needed, it can operate at full load to recharge the batteries and then shut down (and thus maximize efficiency). Unfortunately, all of these batteries and inverters add to the initial expense of the power plant and to the ongoing operations and maintenance costs.

Balance comes in when accounting for the reactive load component of the energy supplied by the power plant to the customer. The electrical loads of a typical industrial or commercial facility (being pumps, motors, compressors, etc.) all operate according to the interaction of the magnetic fields produced in their windings. Some of the electricity used by a piece of such equipment goes to sustaining those magnetic fields, and consequently is not directly available to do work. This is referred to as *reactive power*, and it is proportional to the inductive and/or capacitive nature of the electrical loads receiving power from the on-site generators. One problem I've seen is that the operators of an on-site cogeneration power plant (that is operating in parallel with the local utility) will shift the reactive load to the utility-supplied power. When the intertie circuit breaker opens for whatever reason, thus shifting the power plant to island mode, that reactive supply is instantly shifted to the on-site generators as a step increase in load. This increase in load is on top of any other load the utility was carrying and that also must be picked up by the on-site generators. Having a balance of the reactive load between the utility and the on-site generators can help minimize the impact on the stability of the electrical system when the intertie breaker opens. Clearly, if the reactive load addition is too severe a transient for the generators to handle without tripping off-line, then it defeats the purpose of being able to go into island mode.

Reliability

One fundamental objective of an off-grid power plant, regardless whether it's a stand-alone plant or one that needs to maintain a continuity of power during the intermittent times it's separated from the grid, is reliability. As discussed earlier, having several smaller generators in a modular arrangement is one way to spread the risk that a failure in any one particular generator will severely reduce the overall generation capacity. Having a mixture of power and heat sources available, as well as some energy storage on-site, will also improve the overall reliability of the power plant. If the power plant is completely isolated from the grid, then consideration must be given to the consequences if power to the customer is lost. The cost of this event can be weighed against the cost of additional generation assets and increased on-site energy storage options. Being able to split the customer's loads into vital and nonvital will also help, in that the nonvital loads can be shed when necessary to preserve the continuity of power to the vital loads. Having an on-site emergency diesel generator can also be an option, because it can be used as a last resort—assuming that operating such an engine would be permitted in the first place. Increasing the reliability of any power plant design is nonlinear financially. That is, it will cost a certain amount of extra dollars to increase the reliability of a given power plant from 70% to 80%, but it will cost a substantial amount more to further increase the reliability from 80% to 85%. This is shown qualitatively in figure 4–27.

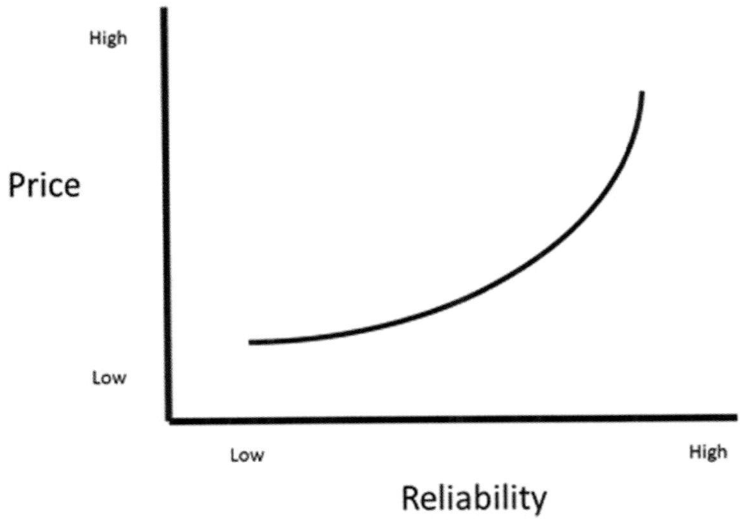

Fig. 4–27. Nonlinear reliability costs

Resynching to the utility

For those power plants that normally operate intertied to the local electric utility and that for whatever reason are now in island mode, what happens when the utility power is restored? A nice feature is the ability to resynchronize the output of the on-site generators with the utility's electricity—including phase rotation, voltage, and frequency—and then reshut the intertie breaker.

Unfortunately, the equipment required in order to enable this transfer back to operations in parallel with the utility comes at a high cost. There are power plants where, to save some money on the initial cost of the power plant, the parties involved chose not to include this resynchronizing equipment. As a result, whenever there was a loss of utility power, the intertie breaker would open and the on-site power plant would operate in island mode just fine; however, when the utility's power was restored to the grid, this particular plant would have to shut everything down, putting the customer's facility in the dark before closing the intertie circuit breaker. As you can imagine, this is rather inconvenient to the customer, and after a few such exercises, the customer will usually pressure the power plant operators to install the automatic reclosing equipment.

Fuel security

Having the best off-grid power plant in the world at your site won't do you any good if it runs out of fuel. If your facility is in a remote location, such as a mine or a ski resort, then it makes sense to address fuel security as part of planning the on-site power plant. While there are natural gas pipelines in most populated areas of the country, those pipelines do not typically extend into remote areas such as mountains. In many cases, natural gas would be the fuel of choice for the power and heat sources of an on-site power plant, but the nearest pipeline is too far away to be useful. The cost of installing a new pipe between the site and the existing pipeline will be an extra burden on the financial aspects of the desired power plant, not to mention the headaches associated with the legal issues (real estate) and the permitting process. For many mountain sites, there is only one navigable access road, and the local electric utility has installed poles along this route to bring electricity to the site. During severe winter storms, falling trees block roads and knock down power lines, and in either case, it is difficult for the utility to get a crew out to the damaged power lines to make the necessary repairs. Also, keep in mind that the more severe the storm is, the more damage it will inflict and the longer will be the time it takes to repair that damage. Such foreseeable delays in the restoration of utility power to the hotels and ski resorts that are electrically downstream of the damaged power lines need to be accounted for in the design of any on-site power plant.

This concept applies to more than just winter storms and mountainous locations. Consider the devastating effects of hurricanes, tornadoes, and other natural disasters that wreak havoc on the conventional infrastructure. Until the roads are cleared, no fuel can be delivered (by truck) to the facility, so some amount of contingency storage of fuel must be a part of the design of the on-site power plant. Also, until electrical power is restored, there will be no way to pump gasoline or diesel into the trucks that would otherwise deliver fuel to the remote location. Any and all natural hazards must be taken into account when analyzing the reliability and risk mitigation of each power plant design/concept. This is true for every power plant location, yet especially so for a site that will be expected to operate in island mode most (or all) of the time.

Ultimately, this brings up the question of which fuel to use, and there are only a few realistic options. The old standard was diesel fuel, because it could be kept in tanks at the site for relatively long periods of time and could be delivered via truck. This is now often a problematic choice, though, because many parts of the country are enacting restrictions on the burning of diesel fuel, so the emissions mitigation requirements may be cost prohibitive. Another viable option is propane, which can also be stored for long terms in tanks and resupplied via truck deliveries. Although there are plenty of original equipment manufacturers selling diesel-powered engine-generator sets, few make units able to run on propane, so equipment selection becomes more of a problem. In any case, one question to answer is just how much on-site storage of the chosen fuel is needed. Using multiple smaller tanks also spreads the risk of a problem with any one tank such as a leak, but this is usually at a higher cost than using fewer larger tanks.

The fuel security situation is usually driven by the anticipated consequences of the expected crises that would put the demands on the power plant in the first place. For example, if the location was a ski resort high in the mountains, any severe winter storm, avalanche, or other catastrophe that makes the access to the facility by road impossible will trap everyone up there. Regardless whether the calamity takes out the main utility-supplied electricity, the resort will need a contingency plan for dealing with all of the trapped people. The ability to supply electricity and heat to such a resort for extended periods of time, under less than ideal circumstances, becomes critical.

High latitudes and geomagnetic storms

Every 11 years, our sun reaches a peak of activity. Considering that the life span of the equipment used in an on-site cogeneration power plant can last 12–20 years or more (depending on how it is treated), I can state with confidence that at some time during the life of the plant, a solar peak will occur. During this time, and during the several years on each side of that peak, solar events can be a problem—although this is more of a problem at higher latitudes than closer to the equator. The problem is that the sun can produce tremendous geomagnetic storms and discharge them out into space, including in our direction. Recall what it takes to generate electricity: a magnetic field, a current-carrying conductor, and relative motion between the two. In this case, the sun is supplying the moving magnetic field. When you take into account structures such as railroad tracks, long-distance pipelines, and even the high voltage power lines that spread our electrical grid across the landscape, you have plenty of current-carrying conductors. If you search the Internet for "space weather," you will find more information about all of this, but my point is that such geomagnetic storms can wreak havoc with power plants. Protection relays can cause trips to happen. Surges in induced currents can damage transformers. Bad things can happen, have happened, and will happen again to power plants and their associated infrastructure. Granted, this is a case where the chances of problems may be low, but they are not zero, and it doesn't take much to protect your power plant from them.

5

Power and Heat Sources

In this chapter, the emphasis will be on the applications of each of the mainstream technologies used in a cogeneration power plant, rather than a detailed description of the inner workings of the equipment. An incredibly wide variety of ways exist to generate power and heat, and they all come with their own set of advantages and disadvantages. There are lots of different manufacturers, design and operational specifications, maintenance requirements, costs, availabilities, service options, and guarantees. In most cases, there will be a trade-off when choosing one option over another, such as to use a naturally aspirated engine (accepting the associated derating) or one with a turbocharger and the required aftercooler (accepting the higher maintenance costs).

My aim is to bring to your attention the factors to consider when using a particular piece of equipment in a cogeneration power plant. I do not mean to imply that any given equipment is inherently superior to any other equipment. For example, regardless of which company made the reciprocating engine, there are specific details to pay attention to when incorporating them in a power plant.

By mainstream technology, I mean the following:

- Reciprocating internal combustion engines (RICEs)
- Combustion turbines
- Microturbines
- Concentrating solar thermal
- Passive solar thermal
- Photovoltaic (PV)
- Biomass
- Biogas
- Geothermal
- Wind
- Hydroelectric
- Fuel cell
- Coal
- Natural gas

- Stirling engines
- Energy storage

How Is It Done?

In all cases of mechanically produced electricity, something is made to rotate (by the prime mover), and with the exception of wind and hydropower, this usually involves the generation of heat via the burning of some kind of fuel. Historically, the heat that was generated was wasted because the only commodity of interest was the electricity itself, so the utility-scale power plants operated at efficiency levels of 30%–35%. In a modern cogeneration power plant, this *waste heat* is captured, and of the British thermal units (BTUs) in each unit of fuel consumed, well over 80% of the energy is put to use.

Consider an internal combustion engine, where the chemical energy released by burning the fuel is transferred into mechanical motion (moving the piston) and into radiant heat (from burning the fuel and friction of the induced movement of the piston, etc.), as well as creating hot gases that are carried away by the exhaust system. Some of that heat is transferred to the engine's coolant, which is circulated through the engine for that purpose; however, some of the heat is transferred to the metal the engine is made of, and this is ultimately radiated to the air surrounding the engine.

There will always be some generated heat that cannot readily be harvested—for example, the heat radiated from the metal block of an engine—but there are people working on it. The benefits of this cogeneration of energy can be reaped by capturing the heat from the combustion of the fuel that was transferred to the coolant, as well as harvesting the heat in the exhaust (typically around 950°F). A basic block diagram of a simple cogeneration system is shown in figure 5–1.

In the basic block diagram (fig. 5–1), natural gas is used as the fuel input to the engine. Cold coolant from the radiator (i.e., a dump radiator) is pumped to the inlet of the engine, then circulated through the engine, before emerging as hot coolant. The exhaust from the combustion process is sent through a catalyst bed to clean up the emissions, usually via an exothermic process, before being passed through an air-to-liquid heat exchanger. The hot coolant from the engine is able to absorb even more heat from the exhaust before going into a liquid-to-liquid heat exchanger, where its heat is transferred to the process fluid. Cold process fluid enters the heat exchanger, absorbs heat from the engine coolant, and exits the heat exchanger at some higher temperature to be used in the facility as process heat. As the engine coolant gives up its heat to the process fluid, its temperature is reduced, so it again becomes cold coolant to be returned to the engine.

The radiator is between the process heat exchanger and the engine as a protective measure. If the heat absorbed by the process fluid is insufficient to

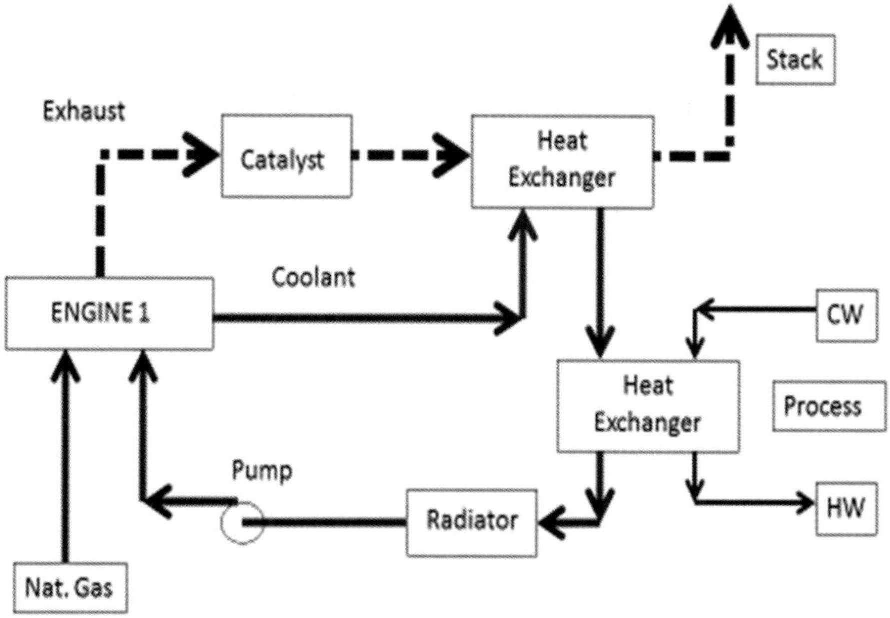

Fig. 5–1. Basic cogeneration power plant block diagram

reduce the temperature of the engine coolant to safe levels for the return trip into the engine, then the excess heat must be rejected, and in this case, it is rejected to the earth's atmosphere—just as the radiator in your car transfers your car's engine heat to the atmosphere. Therefore, that quantity of heat is wasted (dumped). If the facility's process fluid were to stop flowing, then the radiator must be capable of rejecting all of the engine heat and the exhaust heat to protect the engine.

Reciprocating Internal Combustion Engines

In a RICE, stationary engines drive the rotating shaft of an electrical generator. Worldwide, hundreds of manufacturers produce these engines, which are typically sold in pairs as an engine-generator set, or *genset*. The principal manufacturers of RICEs used in the United States are as follows:

- Caterpillar
- Cummins
- Deutz
- Jenbacher
- MWM (Mechanische Werkstätte Mannheim)
- MAN (Maschinenfabrik Augsburg-Nürnberg)
- Guascor

- Waukesha
- Wärtsilä

In general, all of these manufacturers make the same product, and all RICEs work in the same way—after all, the internal combustion engine is not exactly a new technology. The differences between the products are in the details, including the control systems used, maintenance requirements, spare parts availability, specific operating parameters, costs, and delivery times. When selecting which equipment to use in a given power plant, consider how these stationary engines operate, including the following aspects:

- Does the unit use natural aspiration or a turbocharger-aftercooler?
- What is the engine speed (e.g., 1,200 or 1,800 revolutions per minute [rpm])?
- Is the engine driveshaft directly coupled to the generator, or is there a gearbox?
- What is the heat rate (i.e., the amount of fuel needed in the engine to produce 1 kWh from the generator)?
- What is the schedule for top-end and complete overhauls, and what does each kind of overhaul cost?
- What is the periodicity of routine maintenance (e.g., oil changes)?
- What are the emissions levels (e.g., levels of oxides of nitrogen [NO_x])?
- Is it a rich-burn or a lean-burn engine?
- What are the efficiency levels for the engine and the generator?
- Is there an internal water pump?
- Are there internal thermostats? If so, how many are there, and how much flow do they control?

Derating

All of the power and heat sources that use combustion of a fossil fuel for an energy source are subject to *derating*—that is, operating at a maximum output that is less than the unit's nameplate rating. For combustion to occur at the design specifications, there must be a certain amount of fuel and a certain amount of oxygen present at the same time. On the fuel side, consider *pipeline-quality* natural gas; the range of acceptable amounts of methane per cubic foot of that gas will vary from utility to utility and from state to state. Any given natural gas-fueled RICE is designed to run on some minimum methane amount, and it won't run as well (or may not run at all) if the fuel doesn't have sufficient methane in it. The oxygen part of the combustion process is also very straightforward given that combustion cannot occur without the presence of a sufficient amount of oxygen. Consider the amount of oxygen in a cubic foot of air. The higher the ambient temperature is, the less dense the air will be. Hence, the amount of oxygen available for combustion per cubic foot is going to be proportionally less than the initial condition, based on the difference in temperature. Altitude also plays a role, because as altitude increases, the air becomes less dense. Therefore, the elevation of a given site above sea level (or possibly another reference point named in the equipment specifications) will determine the general derating, or reduction from nominal output, that any given RICE will experience; at that location, the RICE will have a lower output in the hot summer months than during the cooler parts of the year.

Using nameplate data—the full rated capacity of the RICE under reference conditions—instead of real-life data in models to predict the performance of any power plant will result in erroneously high expectations that can never be met. Designing a performance model on equipment nameplate data alone may be easy, but the results will be a fantasy. (For a detailed description of a model to determine a power plant's actual performance over the course of a year, see chap. 7.)

Naturally aspirated or turbocharger-aftercooler?

A naturally aspirated RICE is simpler, from an engineering viewpoint, than one with a turbocharger because the air used for combustion is drawn into the engine naturally, as in the engines of most cars. Ambient conditions are thus large factors in the performance of the engine, and elevation (which doesn't usually change) and air temperature (which changes significantly) must both be accounted for. The changes in generator output due to changes in the engine's ability to drive the shaft's rotation, as well as the changes in the fuel efficiency of the engine, are given in the reference specifications of the unit (sometimes in curves), as provided by the original equipment manufacturer (OEM).

One way to mitigate the adverse effects of having variable air densities for combustion is to use a turbocharger (although, as will be shown, this introduces a new set of associated problems). The turbocharger uses a compressor to pressurize the incoming air to the engine by using a turbine that is in the engine's exhaust housing coupled by a common shaft. In this case, the flow of the exhaust is blown through a small turbine, which causes it to rotate (recall blowing on a small propeller mounted on a stick when you were a child), and this rotation spins a compressor at the other end of the same shaft, causing it to bring air into the engine, as shown in figure 5–2.

As the incoming air is compressed, the amount of oxygen per cubic foot of air is increased (desired), but the temperature of the air increases due to friction of the air molecules (not desired). In an engine using spark-induced ignition of the air/fuel mixture, hotter air can induce premature ignition (also known as *knocking*). To minimize this adverse condition and improve efficiency, the inlet air, once compressed, is cooled to again increase its density, giving the engine more air to work with. The incoming compressed air is cooled via a heat exchanger (the aftercooler or intercooler) before it enters the engine. The hot side of this air-to-liquid heat exchanger is the compressed air, and the cold side is usually a water

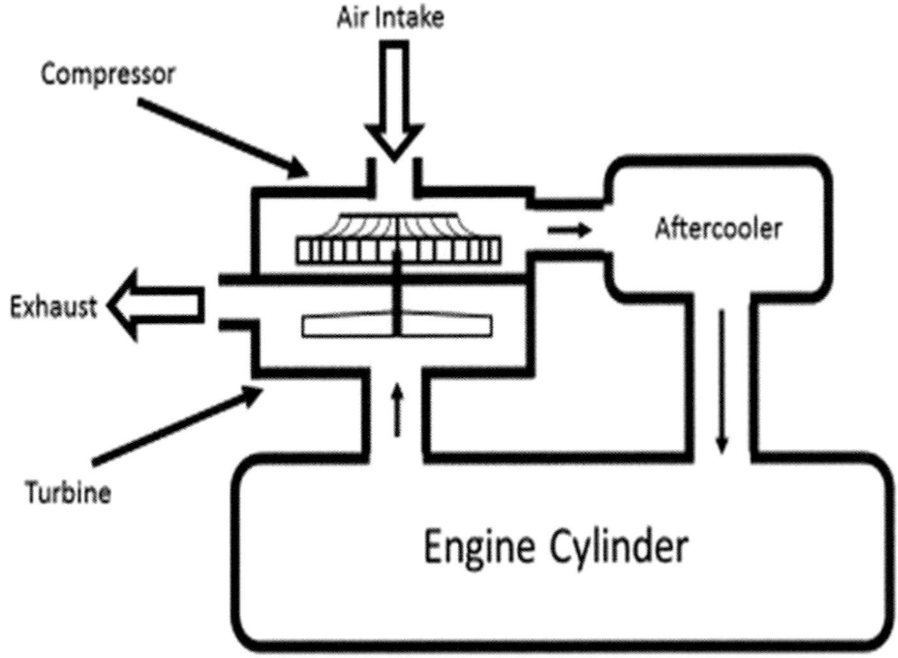

Fig. 5–2. Turbocharger functional diagram

system, or a water/glycol mixture, that absorbs the heat from the inlet air; think of your car's radiator working in reverse, with cold coolant inside and hot air outside. The now-hot aftercooler coolant is then pumped to yet another heat exchanger (or radiator), where the heat is typically rejected to the earth's atmosphere (dumped heat) in the same way your car's radiator transfers heat out of the engine's coolant. One nice feature of radiators is that they don't care which way the heat transfer goes—whether air to liquid or liquid to air, the heat flux will always be from the hot side to the cold side. An air-to-liquid heat exchanger, like the one in a car, is shown in figure 5–3.

A few problems are associated with the turbocharger-aftercooler system, and if these are not addressed during the design phase of the cogeneration power plant project, the consequences can be highly detrimental. First, one limiting factor will be the upper temperature for the coolant supplying the inlet to the aftercooler heat exchanger. This is one of the most common problems I've seen in cogeneration power plants and can severely limit operations in hot months. For example, a common upper temperature limit is 90°F, which means that whatever system is rejecting the heat picked up by the aftercooler must be able to get the liquid's temperature back down to 90°F or below—and if that system is using a dump radiator to reject the aftercooler's heat to the atmosphere, it's not going to be able to do that when the temperature of the ambient air is above about 85°F.

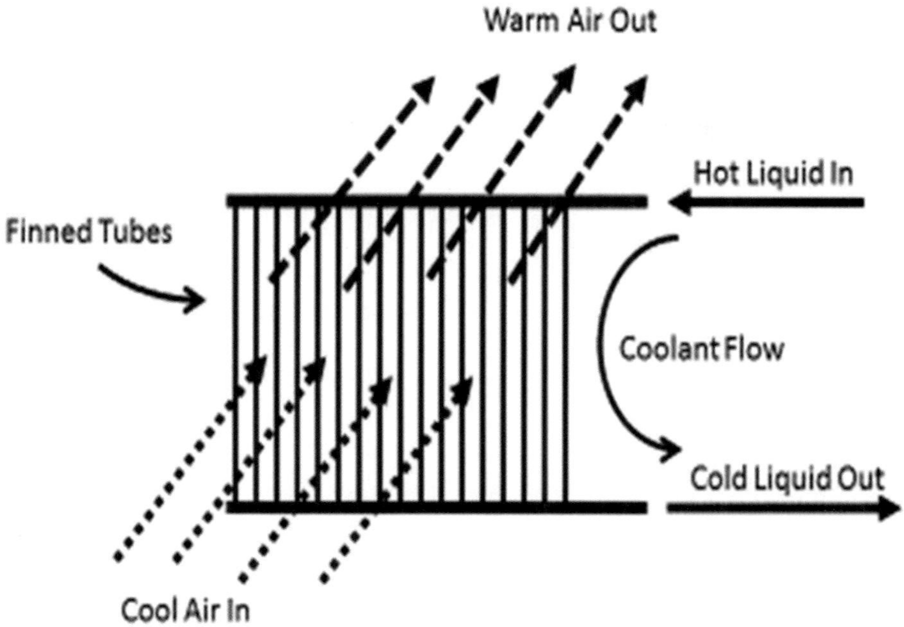

Fig. 5–3. Basic air-to-liquid heat exchanger (radiator)

A critical and pesky detail about heat transfer is that a difference in temperature is required for an exchange of thermal energy to take place. Suppose the temperature of the hot coolant at the outlet of the aftercooler is 105°F, that we need to get that liquid down to 90°F or less before it can be used again, and that we want to use a simple radiator to do so—very similar to the radiator in your car, which rejects heat from your car's engine to the atmosphere. By analogy, if you have a cup of coffee at 105°F that you want to cool to at least 90°F, setting the cup outside when the temperature is over 100°F will not do the job well—the lower the ambient temperature is, the faster the coffee will cool down. Nevertheless, I have seen cogeneration power plants use reciprocating internal combustion engines that had turbocharger-aftercooler systems with the aftercooler inlet temperature limited to 90°F in installations at locations where the high temperature of the day is at least 100°F every day during the summer months—and it remained a mystery to the power plant operators why their engines were not performing up to the manufacturer's specifications.

Consider the situation in figure 5–4. Both the aftercooler and the dump radiator are air-to-liquid heat exchangers, one absorbing heat and the other rejecting it. If the limit on the input to the aftercooler is lower than the ambient temperature of the cool air in, the system will never be able to reject the heat from the output of the aftercooler.

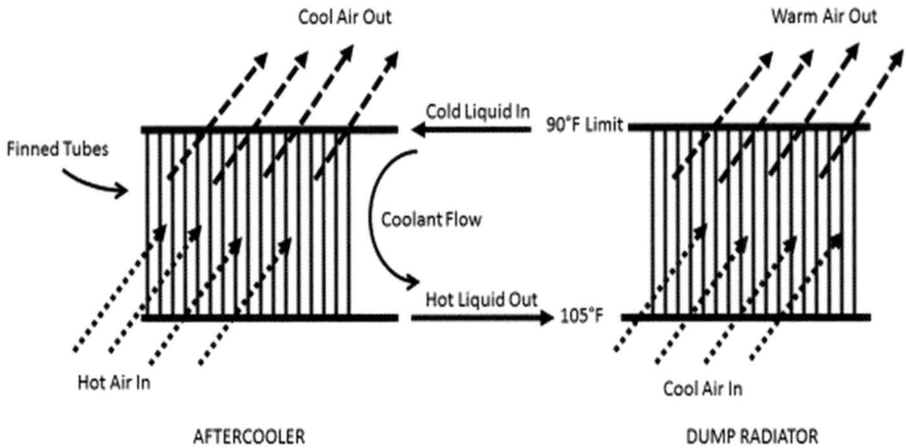

Fig. 5–4. Aftercooler to dump radiator coolant loop

The second big (potential) problem with using an aftercooler is that it is an air-to-liquid heat exchanger, and heat exchangers can fail. In this case, if the tubes carrying the liquid coolant in the aftercooler were to crack or a weld were to fail, then liquid would get into the air side of the heat exchanger—that is, the liquid coolant could then make its way into the combustion areas of the engine, following the same path as the air. If such a leak occurs while the engine is shut down, then water will enter the cylinders instead of air and fuel; since water is incompressible, the next time the engine is started there will be some catastrophic failures of the engine's internal components, and there's typically no way to know whether such a leak has happened. The housing of the aftercooler usually will have one or two ports such that a borescope or similar tool can be inserted into the heat exchanger to look for this sort of problem; however, the typical manufacturer's maintenance plan calls for checking the aftercooler only after significant amounts of operation (e.g., 30,000 run hours). I have seen leaks in aftercooler heat exchangers after only 15,000 run hours. Thus, while a turbocharger-aftercooler system brings some benefits to the power plant, it also brings some design challenges, operational challenges, and maintenance concerns to the facility—all of which need to be weighed in the overall risk management of the power plant.

Third, an operational concern exists when running engine-generators in island mode—namely, *turbo lag*. If you've ever driven a car that had a turbo in it, then you're probably familiar with the experience of turbo lag; that is, you try to accelerate quickly, but only after the turbo picks up does the real acceleration happen. This is because the turbo relies on exhaust flow, and the increased exhaust flow due to an opening engine throttle takes some time; only after this can the compressor side of the turbo add the extra air to the engine, which takes more time still. When the engine-generator is operating in parallel with the local electric utility, the effects of this turbo lag are effectively absorbed by the sharing of power among the generators and the grid. When the engine-generators are operating in island mode, with no electrical support from the grid, even small changes in the electrical load can cause voltage and frequency issues during the transients. Furthermore, if the electrical load transient is severe enough, then the engine-generator unit(s) can be tripped off-line by the protection relays since there are limits on the range of allowed voltage and frequency. If the voltage and/or frequency limits on the system imposed by the local electrical utility via the interconnection requirements (as controlled by the protection relays monitoring the system) are exceeded during the transient, then the relays will send a signal to open the generator output breakers. Thus already being disconnected from the utility, the facility powered by the cogeneration plant will now experience a blackout, and this defeats the purpose of being able to operate the power plant in island mode.

There are three basic ways to get around turbo lag. However, they are each not without their own problems:

- First, and the most obvious, would be to use a naturally aspirated engine instead of a turbocharged one. In part, the decision to use this solution involves weighing the benefits of using a turbocharged engine against the realistic amount of time one expects the power plant to be operating in island mode and the amount of turbo lag each different engine will experience (which is not something the manufacturer will detail on the engine's specification sheet).
- Second, you could use protection relays that are smart enough to know when the plant is operating in island mode as opposed to intertied to the utility (i.e., recognizing when the utility isolation breaker is open or not). These relays can then have a second set of parameters dictating when protection (opening the generator output breaker) needs to happen, and the settings for island mode can be looser than those when operating in parallel with the utility. That said, there are limits to how far the voltage and frequency limits can be expanded for the island-mode electrical system, because most industrial customers use PLCs and other computer-based controls in their facilities—and those control systems are sensitive to the voltage and frequency they are supplied with. It's all well and good for the power plant to remain online after an electrical load change transient, but it doesn't do the customer any good if his equipment trips off-line. Also, the relay must be very fast in recognizing the shift from intertied to island mode and making the transition from the utility-set protection settings to the alternate island-mode settings; otherwise the system may activate protections inadvertently, or protection may not be guaranteed during the transition.
- Third, an additional system could be used to mitigate the problem by stabilizing voltage and frequency on the generator bus while the power plant is in island mode. By analogy, a battery bank can provide a DC voltage source to a DC-to-AC converter (such as an inverter

or motor-generator set) that is independent of the engine-driven generator(s). Here the inverter (or motor-generator set) would stabilize the frequency and voltage on the generator bus long enough to account for the turbo lag time, thus allowing the engine-generator sets to ride through the transient. An inverter-supported island-mode distribution is shown in figure 5–5.

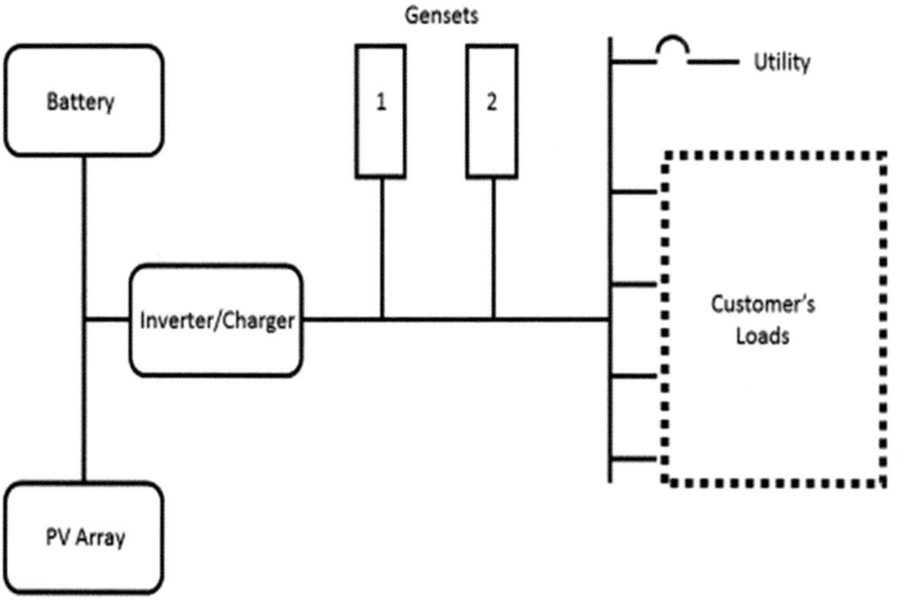

Fig. 5–5. Inverter supported off-grid conceptual power plant

Gearbox?

Another difference in the engines used to drive electrical generators is the speed at which the engine's shaft rotates and whether a gearbox is needed between the engine and generator or whether the two units share the same shaft. In the generation of electricity, the frequency of the alternating voltage is equal to the rotor speed (rpm) times the number of electrical poles used divided by 120 (conversion from minutes to seconds and poles to pole pairs), so a four-pole generator needs to operate at 1,800 rpm to generate a 60 Hz AC voltage. When the engine's design rpm is not a convenient number, a gearbox must be used to change the shaft rpm to what the generator requires. This is similar to the transmission in your car, where you select the speed at which the tires of the car rotate by selecting a gear, thereby changing the rpm of the car's engine. Another analogy is that of a single-speed bicycle, where the rider must pedal at a certain rpm to attain a specific speed down the road (wheel rpm); in the case of an engine-generator set, the engine is designed to operate at a single speed while the generator is also designed to operate at a single speed, and these two rotational speeds are not always equal. This matters in a couple of ways: first, it affects the maintenance requirements of the engine; and second, it affects the maintenance and reliability of the power plant overall.

If the design rpm of the engine is convenient and a gearbox is not required, then the overall operation of the engine-generator set is simplified, and the overall reliability of the power plant is increased. This is true if—and only if—the alignment of the generator is precise. Any misalignment between the generator's stator and the rotor will be a source of problems.

If the rpm of the engine is not convenient and a gearbox must be used, then the overall reliability is decreased owing to the insertion of a *single-point safe* item. This refers to any single item in the power plant where that component's failure will either cause the power plant (or any part of it) to be inoperable or otherwise prevent a safety feature from being operable. In this case, it should be clear that if the gearbox were to fail, then the generator would be unavailable for use, even if the generator itself and the engine remained in perfect working order. In a directly coupled engine-generator set, the failure of either the engine or the generator will cause the entire genset to be unavailable. In a genset with a gearbox, a third point of failure is introduced, and must be accounted for in the reliability analysis of the overall power plant.

Furthermore, in a gearbox (or anywhere else mechanical energy is exchanged), there will be a requirement for both lubrication and cooling to offset the heat that will be generated. If a separate lube oil system is required, with a lube oil cooler, this will obviously introduce additional points of possible failure as well as increased maintenance (changing the oil, cleaning the cooling system's radiator, etc.). If the gearbox is a sealed unit, requiring no external lubrication system, it will still generate heat. If there is no external cooling system for the gearbox, relying instead on direct heat transfer to the atmosphere via a system of louvers in the structure containing the gearbox, there are other concerns that are now introduced. Dust and the oil vapor in the air (from the engine) will combine to form a layer of grime on the outside of the sealed gearbox, reducing the heat transfer ability and thus increasing the temperatures inside the gearbox over time (as the grime layer forms and thickens). The other problem of relying on ambient temperature is that at peak summer temperatures, the gearbox will be operating under high thermal stress, and during any downtime during the winter, the gearbox will equalize with the ambient temperature. For example, if the customer closes his factory for two weeks in late December for the holiday season, as part of a planned shutdown to allow for maintenance of the whole facility, then the lubricant in the gearbox can become congealed. When it comes time to start the power plant back up, the operators need to be aware of the thermal stress limits of the gearbox and make sure they warm up the power plant components per the limits set by the manufacturers of each piece of equipment.

There are other areas of concern regarding the rpm of the given engine, including the impact on the maintenance and overhaul schedule, as well as the engine's response to changes in load. The economic advantage goes to lower-rpm units because the overhaul schedule is typically longer than for higher-rpm engines. With longer times between regular maintenance—and especially between overhauls—the life-cycle costs of operating lower-rpm engines typically become less than those for higher-rpm engines. The simple fact is that engines that operate at a higher speed have proportionally higher wear on their parts and therefore have higher maintenance costs. Lower-speed, larger-bore engines also benefit from a lower surface-area-to-volume ratio, so they experience less heat loss. Medium-speed engines can be more expensive than faster engines, but they also typically operate at a higher efficiency, so trade-off or cost-benefit studies must be done. Two common equipment configurations are shown in figure 5–6.

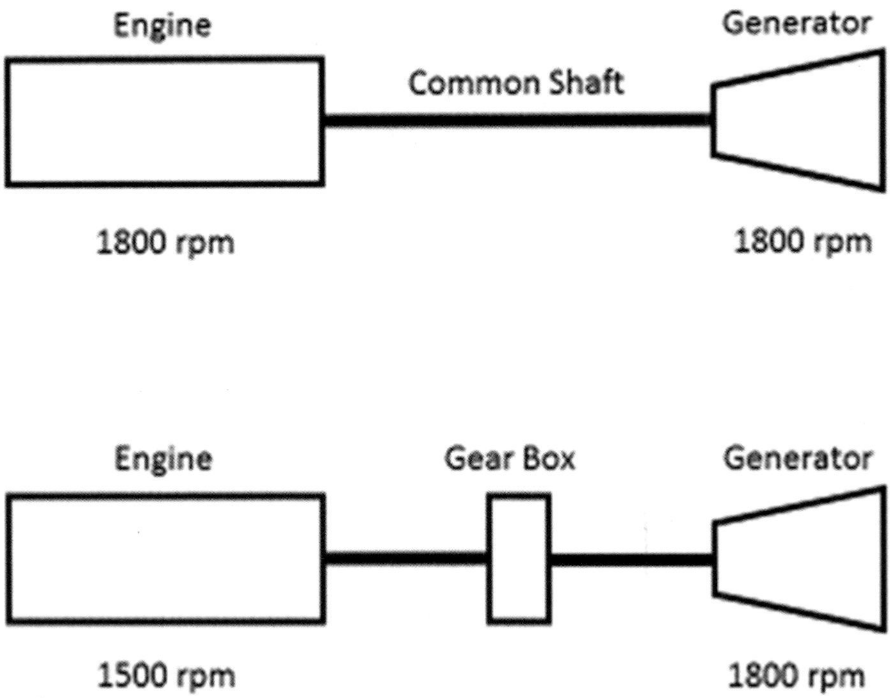

Fig. 5–6. Directly coupled versus gearbox

In general terms, engines that operate at higher speeds are better suited to handle step increases in load, owing to momentum, than are slower-speed engines. This benefit for increases in load is a negative for step decreases in load, for the same reason—namely that the engine will resist the change and as a result may keep the frequency too high for too long during the transient, causing a trip because of overfrequency.

Cost drivers

All things considered, the two largest drivers of the total cost of operating a cogeneration power plant over the lifetime of the equipment are the heat rate and the engine overhauls. The heat rate of the engine (in BTU/kWh) is given in the equipment specifications for the reference conditions of ambient temperature, elevation, and fuel quality, often at 50%, 75%, and 100% engine loads. This is analogous to the fuel efficiency of a new car, where the vehicle's miles per gallon of gasoline is shown for both city and highway driving. At highway speeds, the car's engine is operating at a higher load than at the lower, city-level speeds and is therefore operating more efficiently (hence the better fuel economy). A stationary engine works the same way, but there's more to it than that—a lot more. While the engine in your car may just as easily push you along at 25 mph as at 70 mph, the large stationary engines that drive generators are not like that, and operating them as if they were the same as automobile engines will have expensive consequences.

The engines used in prime power applications (as opposed to backup power generators) are intended for continuous duty at high load levels, including 100%. Each engine manufacturer will have their own set of guidelines on operating the equipment, and one part of that is the management of low-load operation. Most engines can be run continuously from 100% load all the way down to 50% load without any problems (except for the lower fuel economy at lower loads). For operations at less than 50% load, there are time limits imposed before a return to full load conditions is required—and, typically, the requirement is to run the engine at a high load for at least twice as long as it was operating at low load. A typical schedule is shown in table 5–1.

Table 5–1. Low-power engine operating limits

Load	Naturally Aspirated Engine (hours)	Turbocharger-Aftercooler Engine (hours)
Low idle	0.5	1
High idle	1.5	4
10%	4	8
20%	8	24
30%	12	100
40%	24	700
50%	50	>800

The main reason why these large stationary engines cannot run forever at low load levels is cylinder lubrication. When running an engine at very low loads for an extended time, the pressure within the cylinder will be much lower than under nominal conditions, and the cylinder components will end up being overlubricated. For example, consider that piston rings are pushed against the liner wall by cylinder pressure, to create the correct ring-to-liner pressure and, therefore, the correct amount of oil flow; conversely, lower-than-nominal cylinder pressure will have an adverse effect on oil flow.

Other considerations when operating a RICE at low load levels for extended periods of time include higher-than-nominal oil consumption, low intake manifold pressures, lower margin to detonation, reduced transient response (poor engine performance), increased spark plug fouling, and accelerated carbon buildup on valves, cylinder heads, and walls. The problem of carbon buildup is much more pronounced in a naturally aspirated engine than in one using a turbocharger, but a return to high(er)-load operation is necessary in either case. For example, if you operate a turbocharged RICE at 20% load for 24 hours, you should then operate that engine for a minimum of 48 hours at a load of at least 50% (70% for some manufacturers)—and the higher, the better. I remember my father, in the late 1960s, wanting to take his sports car out onto the highway on the weekend (after a week of city driving) and "really open her up" in order to "blow all the carbon out"; now I know what he was talking about.

Modeling

When modeling the performance of any given engine, once the loading of the generator is found and then converted into a percentage (i.e., the engine loading), the heat rate (or fuel efficiency) can be calculated for every load level using a very simple equation. (Heat rate will also be discussed in chaps. 6 and 7.) My point here is that the average amount of fuel consumed during each 15-minute interval can be predicted with a fair degree of precision. Therefore, the total fuel consumption of the engine can be estimated for any particular time period, and then the cost of that fuel can be determined—and when this higher-precision cost estimate is used, instead of the full-power nameplate data, a more realistic assessment can be made of the true costs associated with operating the genset. This difference between reality and nameplate expectations is usually more severe than the engine equipment salespeople would like you to think. This is analogous to the realization, when operating a new car, that the combined highway and city driving gas mileage is actually far less than what was promised for driving only on the highway; intellectually, you knew it would be less, but just how much less may be a bit of a shock.

Overhauls

Each engine manufacturer will have their own scheduled periodicity for performing an overhaul of the engine, and in general with RICEs there are two kinds:

the *top end* (every 1–3 years) and the *bottom end* (usually twice the periodicity of the top end, or every 2–6 years). A top-end overhaul typically involves installing new (or refurbished) cylinder heads and turbos, whereas a bottom-end overhaul involves replacing the pistons, cams, main shaft, bearings, and other parts, as well as all the top-end work. There are also different sets of logistics that vary from manufacturer to manufacturer, as well as location to location across the country, with regard to how exactly the overhauls are done. In general, there are three variations:

- First, and the fastest, is for the engine requiring the overhaul to be removed from the site, and a new (or refurbished) unit is immediately installed in its place. The engine is craned onto a truck and taken back to the manufacturer's facility for the actual overhaul, and then it becomes a standby unit to repeat the process when it is next due. This engine-swap method has the advantage of being fast to execute, so that the power plant is back up and running in very short order, thus restoring the revenue stream and benefits of using cogeneration over the local electrical utility to the customer. A potential disadvantage is that the permitting for the emissions coming from the exhaust of the engine is usually tied to a particular engine, by serial number, such that when swapping in a "new" engine, the owner must file for a new permit to operate with the local air pollution control district—which is not usually quick to obtain.
- The second variation of overhaul logistics is similar, in that the existing engine is removed from the power plant and trucked to the manufacturer's facility for the work, but that same engine is then reinserted back into its place in the power plant once the work is complete. This has the advantage of removing any administrative burdens with the local air pollution control district, but it carries the disadvantage of taking some time—the engine may be out of service for several (usually two to four) weeks before being returned to service.
- The third variation is where the engine technicians come to the power plant every day, for 12 hours or more each day, and work on the engine in situ for as long as it takes to complete the overhaul. The advantage of this is not having to use a crane (and all the associated logistics of getting a crane on site, twice); however, it also has the disadvantage of taking longer (typically) and the possibility of encountering something wrong in the engine whereby it cannot be repaired on site, because a special tool is needed (and thus another trip to the shop by one of the mechanics) or a part is not in the warehouse and must be shipped in (adding delays). For coordination of work with the equipment manufacturer and their technicians, a cost-benefit analysis can be done to manage the risk by minimizing downtime and costs.

Oil changes

Another parallel to the engine in your car is that stationary RICEs use oil as a lubricant; further, as in your car, there will be an oil sump, an oil pump, a filter, and a heat transfer mechanism (usually a radiator). The type and quantity of oil for the genset engine will vary from manufacturer to manufacturer, as will the periodicity of when an oil change is required. Some gensets will include extra on-site oil storage, making the entire oil system a large-volume operation, thus extending the time between oil changes (but requiring much more new oil with each changeout). Other oil-based systems include a degree of on-site oil purification, usually using some kind of centrifuge, to extend the time between oil changes. From an O&M perspective, oil changes will require a certain amount of engine downtime to pump out the old oil, pump in the new oil, and change out the oil filters. An obvious hazard here is the possibility of an oil spill, and local hazardous waste (hazmat) ordinances must be complied with, including fire prevention measures, and spill containment.

Clearly, since all engine downtime is to be minimized, the engines with more oil changes per year are not as desirable as those engines that have less frequent maintenance requirements. The choice of which engine to use as an engine-generator package must address the amount of downtime and the associated costs of being off-line as well as the costs of the consumables (oil and filters). Do not overlook that when an engine is down, even for planned maintenance like an oil change, the rest of the power plant is more vulnerable to an unanticipated problem, and overall plant reliability is reduced.

Reliability trees

Let's say you just bought a reciprocating engine-generator set, and the manufacturer states that there's a 10% chance that the unit will go off-line (from any one of all the possible causes) at any time. You can build a diagram showing the relative probability that the unit will be available (90%) for operation (or not), as shown in figure 5–7.

If we buy a second unit, of the exact same kind of genset, we can add the additional probabilities to the diagram, as shown in figure 5–8. The probability that both gensets are available is now 81%, whereas the probability that both units are off-line is now only 1%. The probability of having only one unit available (either engine) is 18%. If we keep adding additional units to the site, the probability that all of them will be available at the same time will continue to decrease, but the probability that all of them will be unavailable will also decrease.

Since "or" probabilities are added whereas "and" probabilities are multiplied, a chart can be made showing the various chances of each possible configuration of the equipment (e.g., gensets 1 and 3 are available but genset 2 is not). Consider having three identical units. The probability that all three units will be available for use is reduced to 73% (i.e., 0.9 × 0.9 × 0.9). At the same time, the probability that all three units will be simultaneously unavailable for use is reduced to 0.1% (i.e., 0.1 × 0.1 × 0.1).

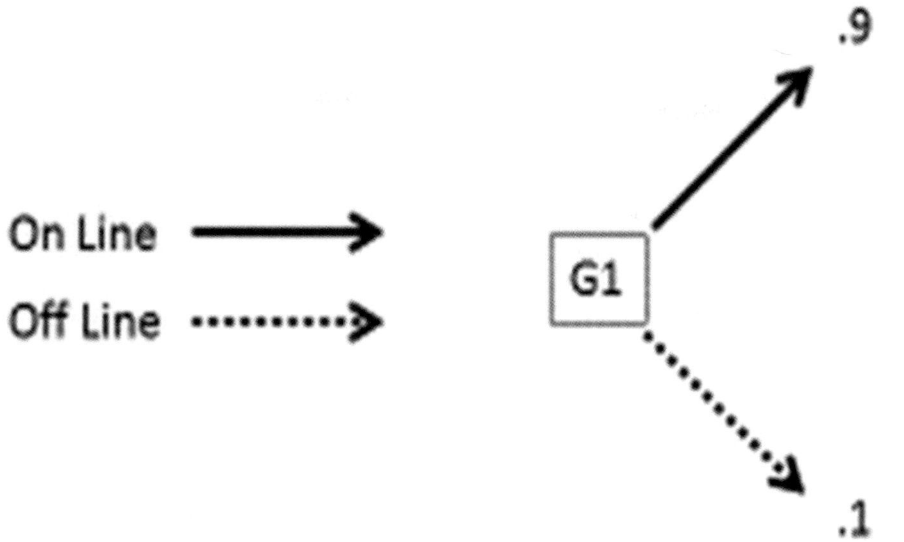

Fig. 5–7. Single-engine reliability tree diagram

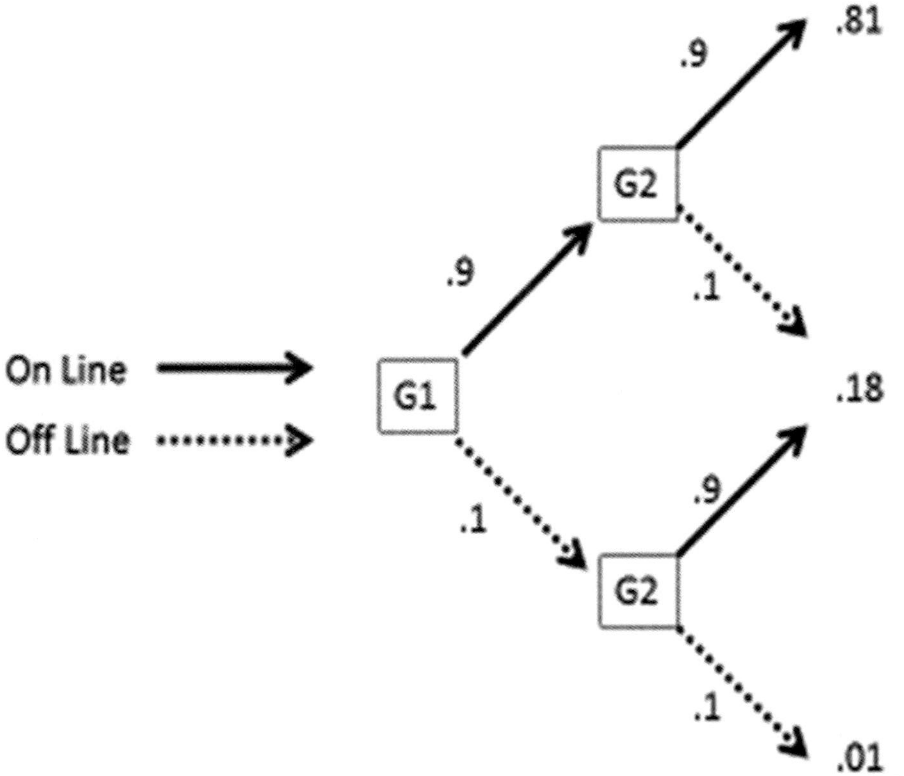

Fig. 5–8. Two-engine reliability tree diagram

The idea is to drive the probability that all of the electrical generators are unavailable as low as possible while minimizing the overall project costs. Having more smaller generators is more expensive than having fewer larger units, but this is offset by the increased probability that at least some of the assets will be available for use. A key point to consider is the cost to the customer when all the power is out versus the costs associated with the initial power plant equipment (and the lifetime maintenance costs). Similarly, it may not take a complete power outage to hurt the customer financially; rather, it could be as simple as not having enough power to run half of her facility/factory. In these cases, the minimum on-site generation capacity needs to be determined, then the acceptable probability of not being able to supply that capacity.

Emissions levels

Included in the specifications for each engine will be the amounts of the various pollutants in the exhaust, such as oxides of nitrogen and sulfur (NO_x and SO_x, respectively), carbon monoxide (CO), and carbon dioxide (CO_2). The relative importance of these emissions is based on location, because some states have much more restrictive limits on the amount of emissions than others, and there are variations within states. In many cases, the acceptable levels of emissions are based on the *best available control technology* (BACT), and these standards are loosely based on whatever the U.S. Environmental Protection Agency (EPA) is currently considering to be the best. There are other triggers, including the Clean Air Act and EPA Title V permitting issues, especially in states with stringent regulations; for example, California has imposed a *carbon tax* on any equipment that emits more than 25,000 metric tons of CO_2 equivalent per year. For RICEs, that imposes a de facto upper size limit of about 5–6 MW of electrical generation capacity, above which the impact of having to pay carbon taxes will have a very negative effect on the financial performance of the power plant. This carbon tax, where it is imposed, creates a power plant-sizing dead zone where facilities smaller than about 5.5 MW are financially viable. It's not until the economy of scale comes into play before larger plants become viable—that is, they need to be large enough that the financial impact of the carbon tax can be absorbed and the facility still remain profitable enough to be built. The net effect is that no one is going to be building fossil fuel-burning power plants in places like California, cogeneration or otherwise, that operate in the range of about 6–35 MW generation capacity (unless they somehow get a reprieve from the carbon tax by the state government).

Once the net emissions from each RICE used in the power plant have been calculated, a plan can be conceived to deal with the magnitude of the emissions problem. There are a variety of catalyst-based cleanup measures available via selective catalyst reduction (SCR) or nonselective catalyst reduction (NSCR) depending on whether the RICE is a lean-burn or rich-burn engine. There are also separate units for cleaning up the CO in the engine's exhaust, most often converting the CO into CO_2, and these are usually upstream of the SCR in the exhaust path.

Rich or lean burn?

The equipment used to clean the emissions from engines depends on the operating nature of the engine—specifically, the air/fuel mixture that is used and, therefore, the temperatures reached during the combustion process. Stationary engines that burn a gas (as opposed to a liquid fuel) for combustion generally are either *lean burn* or *rich burn*. The terms come from the amounts of fuel and air in the combustion chamber, relative to the amount of fuel—that is, a rich-burn engine operates at a higher fuel-to-air ratio than a lean-burn engine does. While the specific differences between the two types of engine will vary with the given manufacturer, certain generalities describe the differences in load carrying ability, operational flexibility, and derating effects.

In general, lean-burn engines can operate at higher loads than rich-burn engines, and as such, larger units will tend to be lean-burn engines while smaller units can be either type. Rich-burn engines have an advantage over lean-burn engines in that the cleanup of the exhaust is simpler. For the combustion of the fuel to be complete, a certain minimum volume of air must be present, and rich-burn engines use just enough air to burn all of the fuel—that is, they are *stoichiometric*. Combustion without excess air allows for the use of an NSCR catalyst as a single treatment to clean the exhaust without the need for a separate carbon monoxide treatment or a source of ammonia or urea. There are limits, however, in the kind of gas that can be used as the fuel for rich-burn applications. Rich-burn engines can operate on a variety of natural gas fuels, but other sources of gas fuels (e.g., biogas) cannot be used because some of the emissions from the combustion of alternative fuels contain compounds that will damage (poison) the NSCR catalyst material.

In contrast, lean-burn engines use up to twice the amount of air needed for complete combustion of the gas, which reduces the temperatures within the combustion chamber, yielding a couple advantages over the higher-temperature combustion—in particular, emitting lower levels of pollutants. However, the advantages of lean-burn engines come at a cost, because a more complex emissions cleanup system is required. Having a lower temperature at the point of combustion reduces the amount of NO_x produced and allows operating at higher loads owing to reduction of the probability of engine knock and improving the fuel economy. These advantages are offset by the disadvantage of needing a separate oxidation catalyst to first convert the CO in the exhaust into CO_2, after which an SCR catalyst cleans up the exhaust, which requires a source of ammonia in the chemical cleanup reactions. This reactant is commonly injected into the exhaust stream prior to a mixing chamber in the exhaust path, where it can be evenly distributed in the exhaust before entering the SCR itself.

In general, there are two sources for this reactant: urea or ammonia proper (either liquid or anhydrous ammonia). Having large tanks of liquid ammonia at a power plant poses safety concerns, including control of the liquid if the tank should fail, as well as of the ammonia vapor that would be released in the event of a major spill. An alternative is to use bottles of anhydrous ammonia, which are smaller and safer than having a big tank of liquid ammonia, and the injection rates are typically small enough that a standard U.S. Department of Transportation (DOT)-rated bottle will last a considerable while before needing replacement. In either case, the ammonia will react with the nitric oxide (NO) in the exhaust to form nitrogen gas (N_2), along with water vapor, as it enters the SCR. Liquid urea is easier to use than liquid ammonia but still requires spill containment measures and poses other problems—in particular, as a corrosive liquid, it will eat the impellers of the pumps used to get the urea into the exhaust stream. Also, any threaded connection in the stainless steel piping from the urea tank to the exhaust header can experience a leak that results in the formation of white crystals (which make it easy to spot the leak). The urea will also react with the NO in the exhaust stream, forming N_2, water vapor, and CO_2.

When urea is used and the reagent injection system is off-line (e.g., to replace the impellors in the pumps again), the urea can crystalize in the injection lances in the exhaust header. These are more like a shower nozzle than a garden hose, with very small flow paths to disperse the urea into the exhaust as a fine spray or mist. Nevertheless, when you turn off the flow, there's still urea between the pump discharge and the injection nozzles, as there would be water in a hose downstream from a spigot; thus, when the engine is shut down (for whatever reason) and the exhaust header cools off, the urea that's still in the injection lances can crystalize, clogging the nozzles. This means that when the engine is restarted and the emissions control system is brought back into service, it will still be off-line, because even though the urea pump is running, the reagent is not getting into the exhaust stream.

If the emissions levels coming out of the exhaust stack are not immediately monitored, the power plant could operate without emissions mitigation (i.e., out of compliance). Many operating permits require sampling the postcatalyst exhaust only once a month. However, if the technicians from the air-quality board make a spot check of the emissions levels, a hefty fine could ensue. From an O&M perspective, when urea is used, it's critical to clean the entire emissions control system any time it's been off-line for more than a day, or at least perform an inspection of the injection lances before starting the engine. Having a spare set of these lances is handy, because they can be swapped out with the currently installed ones, getting the power plant back online faster (and the used lances can then be cleaned at leisure). Also, if the power plant does not have a *continuous emissions monitoring* system, an easy recommendation is to take a sample of the emissions going up the exhaust stack as soon as the catalyst is completely up to operating temperature.

Engine and generator efficiency levels

There are many different makes and models of both engines and generators, and each will have its own rated efficiency (usually given as a percentage) for both electrical and thermal efficiency. The thermal efficiency is the amount of useful heat that can be harvested from the unit compared to the thermal energy of the fuel that is consumed. While the quest for equipment of the highest operational efficiency is often a driving force in the selection of what equipment to use in a power plant, there are cases where opting for a lower-efficiency unit is the better choice over the life of the facility (owing to better thermal harvesting). On the electrical side, the given generator will have an overall efficiency based on the amount of engine horsepower that is converted into electrical energy. Generator losses include the following:

- Field, exciter, and armature losses owing to phase currents, phase resistances, field currents, and field resistances
- Construction element losses owing to iron characteristics, hysteresis, and eddy currents caused by flux variations in the core
- Friction and windage losses (associated with moving the rotor through the air)

Similarly, there are differences in the mechanical losses from one engine to another, but they are all based on how the fuel is burned to make the main shaft rotate by converting linear motion (of the piston) into rotational motion (via the cams). These can be grouped as follows:

- Friction, from all moving parts against one another, and the condition of the lubricating oils, etc.
- Mechanical, from the size of the bore, stroke, angle of the cam shaft, materials used (weight), etc.
- Fuel handling system, including turbochargers and aftercoolers
- Temperature control (how well heat is transferred) within the engine

The efficiency of genset units is also affected by the quality of the fuel being used—ultimately, the driving power of the whole system is based on the ignition and combustion of something. Units designed to run on biogas, landfill gas, or sewer or wastewater treatment gas will have electrical efficiencies of 25%–43% and thermal efficiencies of 38%–64%, and these are typically offset. That is, if a genset has a high electrical efficiency, it typically has a lower thermal efficiency, and vice versa. For units using natural gas, you can find electrical efficiencies of 30%–47% and thermal efficiencies of 39%–64%, with total efficiency numbers in the 73%–95% range.

Internal water pump?

Not every engine comes with an internal water pump for the main engine coolant loop, and those without one will obviously need an external pumping system. This takes up more space and adds some parasitic electrical load, but allows for increased reliability by using two pumps in parallel (with one as an installed spare). For those engines that do have an internal water pump, you can do the engine a favor by making the coolant path as short as possible. I am in favor of running the hot engine coolant directly from the engine outlet to a heat exchanger, where the heat can be harvested, and then to a second heat exchanger (e.g., a dump radiator) for engine protection in the event the primary heat sink is lost. This configuration is shown in figure 5–9.

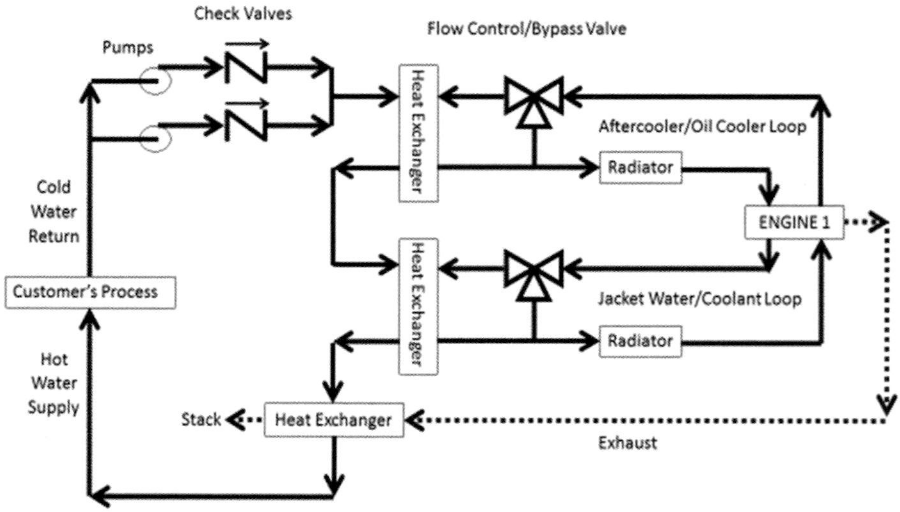

Fig. 5–9. Engine coolant flow-path design concept

In both the aftercooler/oil cooler (commonly combined in reciprocating engines) and the jacket water (main engine coolant) loops, the hot liquid is pumped out of the engine and straight into a flow-control valve. The controls on this valve allow the liquid to flow to the heat exchanger for harvesting, or take a bypass path directly to the dump radiator. The amount of fluid flow in each path and the speed of the fan blowing air onto the radiator are controlled by a PLC or computer controls on the basis of the fluid temperatures. The temperature of the fluids in the systems is measured at the entry and exit of all the pieces of equipment, so fine control over the heat balance can be achieved. These temperature data can be stored for later review in a database. It is handy to periodically review the trends in the data to anticipate problems, as well as to examine the conditions just prior to when something happened (e.g., an engine overheating).

I have seen a power plant where the engine's internal water pump was supposed to provide coolant flow through a complex system of parallel heat exchangers that were in the exhaust path (to heat the engine's coolant some more) before going to a heat exchanger for harvesting of the thermal energy, and then through yet another heat exchanger (the dump radiator) before returning to the engine—with several possible bypass paths as well. Oddly enough, the heat transfer amounts seen in reality never came anywhere near what was envisioned by the designing engineer—partly owing to the actions of the thermostats inside the engine (that were not taken into account) and partly owing to insufficient pumping power of the engine's internal water pump.

Internal thermostats: How many, and how much flow do they control?

Engine manufacturers try to make their engines idiotproof whenever they can, and one way of doing this is to use thermostats inside the engine to control the coolant flow such that very cold coolant will not come into contact with very hot engine parts. The number of internal thermostats, as well as the fraction of coolant flow they control, will vary from one engine to the next. The key point is that while coolant is recirculating internally (so it can warm up), the amount of coolant flow outside the engine will be reduced by some fraction. The result will be a reduction in flow through the various heat exchangers on the hot coolant side, so if the flow of the fluid in the heat sink side of the heat exchangers remains unchanged, then the heat balance will not be ideal. If the heat sink is not a variable and/or if bypass flow paths are not provided, then a thermal runaway condition can occur (as described in the fourth case in chap. 7).

Costs

There are so many different manufacturers of reciprocating internal combustion engine-generator sets around the planet and such a wide range of features that shopping for one based on price alone is a challenge. This is similar to comparison shopping for a new car, where a generic sedan will have hundreds of options to choose from. In general terms, there is an economy of scale at work, where smaller units will cost more on a dollars-per-kilowatt basis. For units under 5 MW of electrical output, the range is from about $1,500 to over $5,000/kW. For larger units, this figure drops to about $1,100–$2,500/kW. Again, the cost of having additional generation assets on-site must be weighed against the costs associated with a loss of power to the customer, whether to the whole facility or just a portion thereof.

Combustion Turbines

Combustion turbines are essentially modified jet aircraft engines, and have been around since the 1940s. They operate at fairly high rotational speeds and discharge a significant amount of high-temperature exhaust. These are typically large in size, and their generators range from about 3 to 250 MW (with a few even larger) and can be used just for electrical power generation (which would be very wasteful) or they are combined with a heat recovery system for cogeneration. They operate at fairly high efficiencies and have a clean burn of the fuel, which limits the pollutants in the exhaust; in fact, when using natural gas as the fuel, combustion turbines emit the least NO_x per kilowatt-hour of generation. A typical configuration is shown in figure 5–10.

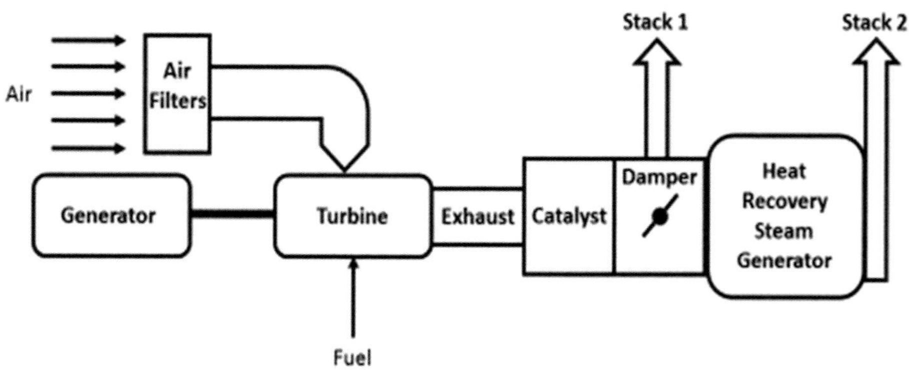

Fig. 5–10. Combustion turbine configuration

Advantages of combustion turbines

High-quality exhaust. Combustion turbines generate massive amounts of high-temperature exhaust, making them ideal for heat recovery, and the exhaust typically passes through a catalyst system to clean it up before it is used. Even though it's cleaner exhaust than that from a reciprocating engine, it still needs a bit of cleaning. After the catalyst, there needs to be some control over the flow of the exhaust, and this is where many combustion turbine-based power plants first get into trouble.

It is convenient to use a damper system to direct the clean exhaust either up and out of the first exhaust stack (wasted heat) or to the heat recovery steam generator (HRSG). I have seen this damper positioned to control the steam pressure in the HRSG, since steam demand is often variable. The problem with this method is that the system used to position the damper gets cycled a lot, since steam pressure is always in flux, so even pneumatic damper positioning systems can fail quickly—so the failure of a $500 part can require the whole power plant to be shut down. A simple idea on paper can end up being an O&M nightmare if redundant systems are not used.

Low maintenance. Many combustion turbines use relatively large reservoirs for their lube oil systems, and many also contain an on-skid lube oil purification system, such that oil changes are infrequent. The inlet air filtration systems typically include a back-pressure cleaning system; when the differential pressure across the air filters reaches some threshold of pain, a puff of high-pressure air is applied from the turbine side of the filter, thus blowing any accumulated dust and debris off the face of the air filter. Most advanced turbines use a predictive modeling system, together with detailed monitoring of the unit, to determine the maintenance requirements (as compared to a system based on a calendar or the number of run hours elapsed). Even the annual borescope inspection requirement can be extended to the two-year point if all the other parameters are within specification, with full overhauls as long as 50,000 run hours.

Relatively quick start time. When a turbine is started, it is described as going from *cold-iron* to *flame-on* conditions. During this warm-up process, there is a very real danger of causing excessive thermal stress on the blades of the turbine during the warm-up of the equipment. For a large steam turbine, the warm-up time can be up to an hour as the blades farthest from the steam inlet get the chance to heat up—most of the heat in the steam is consumed by the cold blades closest to the steam inlet. Steam is slowly introduced in small doses at first, then in larger amounts as the blade temperatures equalize.

The situation is similar when a combustion turbine is used; however, the heat source is a lot stronger than in a steam turbine, and there's really no gentle way to go from cold-iron to flame-on conditions. There is still a prescribed warm-up period by each combustion turbine manufacturer; this usually consists of slow ramping increases in load, followed by plateaus of specific times to allow the thermal stresses to equalize before proceeding to full-load operations. The total warm-up time for the combustion turbine can be as short as 15 minutes.

Containerized. Although combustion turbines can operate in the open, exposed to the elements, most units are housed inside either a building or their own container for protection. There are pros and cons to putting any heat source inside a container, and different sets of concerns pertain depending on the type of heat source; for example, I will rarely recommend a container for a reciprocating engine, but I do think they're a good idea for most applications using combustion turbines. The inside of the container stays cleaner and exterior noise is reduced, among other benefits.

Low emissions. Combustion turbines can have NO_x emissions ranging from about 25 parts per million by volume (ppmv) for smaller units to about 15 ppmv for larger ones, with CO levels of 50 to 25 ppmv for smaller and larger units, respectively. One way to reduce emissions that is available to combustion turbines but not really an option for most other combustion equipment is to inject demineralized water (or steam) into the high-temperature flame zone in the combustion chamber. This can lessen the severity of any hot spots and thus reduce the formation of NO_x by as much as 60%. This is the same way a lean-burn RICE uses excess air to lower the temperature in the combustion chamber and thereby decrease NO_x formation.

Use of duct burners. This is the second way some combustion turbine-based power plants get into trouble. The exhaust of a large combustion turbine will be fairly high in excess oxygen, on the order of about 15%; this allows for the ability to insert a post-turbine combustion process to increase the thermal energy used to generate steam in the HRSG. This is most often done using *duct burners,* which are nothing more than lances injecting fuel (usually natural gas) and a spark source directly in the exhaust plenum of the turbine. One big problem with duct burners, especially in states with strict emissions regulations (e.g., California), is that this combustion is taking place downstream of the emissions control systems. The raw exhaust of the duct burners, together with the cleansed exhaust of the turbine, goes straight out the exhaust stack into the atmosphere. Even though a second SCR unit could hypothetically be installed on the second exhaust stack, that would considerably add to the construction costs of the power plant, as well as increasing the costs of maintenance and testing of the facility. (A case study describing a combustion turbine power plant that failed and how the use of untreated duct burners contributed to the failure is given in chap. 7.)

Disadvantages of combustion turbines

Not good to operate at low load. Reduction of the output of a gas-fired turbine is achieved by lowering the inlet temperature, which also reduces the operating efficiency and increases emissions and the heat rate. Although combustion turbines can be operated in the 80%–100% of full load range, there is a nonlinear decline in efficiency as load is reduced below that. Each turbine manufacturer will have their own set of curves describing these effects, but in general you can expect a serious change in operating parameters, as described in table 5–2.

Table 5–2. Combustion turbine low-load operations

Load	Output (kW)	Fuel Flow (million BTU/h)	Heat Rate (BTU/kWh)	Thermal Efficiency (%)	Exhaust Flow (lbm/h)	Exhaust Temperature (°F)
100%	4,116	49.24	11,963	28.52	139,482	960
90%	3,705	45.10	12,173	28.03	134,528	922
80%	3,293	41.50	12,601	27.08	122,801	941
70%	2,881	38.68	13,423	25.42	113,590	966
60%	2,470	35.90	14,536	23.47	104,529	993
50%	2,058	32.89	15,980	21.35	94,680	1,024

Table 5–2 shows clearly that even though combustion turbines have their advantages, they are really not suited to facilities where their load will be less than 80% except for brief periods of time. If they are used in places with large variations in load, the performance model needs to include all of the negative effects of low-load operation to accurately predict the financial impact on the viability of the power plant.

Sensitive to dust, need lots of air. Combustion turbines require significant amounts of air—if the exhaust flow is on the order of 150,000 pounds per hour, think about how much air needs to go into the front end of the unit. This should immediately bring to mind a couple of points: first, the air needs to be relatively clean; and second, if the unit is inside a building, then that building needs sufficient ventilation to prevent drawing a vacuum.

Even though combustion turbines will always have an effective filter system on inlet air, the operating environment will have a direct impact on how fast those filters will get clogged to the point that the differential pressure across the filter becomes too high. Many of these air filter systems will have a supply of high-pressure air on the turbine side of the filter that is used to back-flow the air filter with a puff, thus blowing all the collected dust and debris off the face of the air filter and restoring it to nominal operation. In the case where these air filters are exposed to the elements, this back-pressure method is of limited use; once the dust and dirt get rained on, it can harden like adobe or cement, and mechanical cleaning of the air filter is then required.

If the turbine is inside a building (to provide a more controlled environment), then the ventilation systems used in that structure must be given serious thought. In addition to the amount of air that must enter the building to prevent a vacuum from being drawn, the environment outside the building must be taken into consideration. In particular, if it is situated in a dusty or industrial environment, then whatever filters are on the building will need attention.

One facility I know takes wood debris and pulverizes it into various sizes of sawdust before gluing the material together under pressure to make fiberboard (similar to plywood). This facility used a fair amount of power and steam, so an on-site cogeneration power plant seemed to be a good fit. One of the problems they encountered was that there was sawdust everywhere, and consequently the building housing the turbine never had sufficient ventilation. Moreover, there was always a vacuum when the turbine was operating, making it a challenge to open the doors to the building. Worse yet, every time one of the doors to the building was opened, the vacuum would suck in large amounts of sawdust, which would get trapped by the air filter on the turbine. The problem was that when the turbine's air filter system would try to clean the filter with a puff of air, the dust and dirt would fall to the floor of the building and, if not cleaned up immediately, would just get sucked right back onto the filter again.

Failures can be catastrophic. Combustion turbines can have a wide range of failure modes (as can any other large piece of equipment), but there are many failure modes peculiar to turbines that end up causing massive amounts of damage. Specifically, there are several kinds of blade failure—such as incompatible differential thermal expansion of the metals, welding failures, or fatigue that ends up creating a small piece of metal that is free to move. Given the number of fixed components and moving blades (and the speed of rotation), a loose bit of metal could easily cause a cascade of shrapnel as the turbine quickly eats itself. Therefore, even though the probability of failure may be small, the consequences of failure are severe.

Loss of lubrication can also be a problem because so much spinning mass can take a while to come to a full stop. Thus, redundant lube oil systems are a minimum requirement to protect the asset, including systems that can provide lubrication in the event of a loss of all electrical power. Backup power supplies for lube oil systems, such as batteries or compressed air bladders in storage tanks, are examples of contingencies to keep in mind when evaluating the use of a combustion turbine. When these systems are employed, there must also be an understanding of their added maintenance and testing needs, as well as their impact on overall system reliability.

Often used with dampers to control HRSG pressure. In the typical turbine-based cogeneration power plant layout, the control of the amount of heat going into the HRSG—and thus regulating the steam pressure—is achieved by adjusting the amount of exhaust that is directed up an exhaust stack prior to reaching the HRSG (as shown in fig. 4–10). The actuators for this flow damper can be a very weak point in the overall design. Unless the nominal pressure-temperature band for operating the HRSG is fairly large, the damper positioning control system can end up *hunting*—that is, continuously adjusting in an effort to operate at the set point (like chasing a moving target). Hunting leads to cycling far more than originally anticipated.

Consider the thermostat in your home: if you set it to maintain your home at 68°F, then the system will cycle on and off as the limits of the set point are reached. The range of values where the system does not respond is known as the *deadband*. If the deadband has a width of 1°F and the system is set to heat, then the threshold will be reached when the temperature sensed by the thermostat reaches 67.5°F, and the system will energize the heating element; similarly, once the temperature is sensed to have risen to 68.5°F, the system will de-energize the heating element. With these types of control systems, whether it be for the temperature of your home or the temperature (and, hence, the pressure) within an HRSG, it should be clear that the smaller the control system deadband is, the more the system will cycle on and off. Conversely, even if the deadband is relatively large, if the variable under control has large swings in value above or below the range of the deadband, the more the system will cycle. Such swings and responsive adjustments exemplify hunting.

In the case of a turbine exhaust-flow damper, subtle changes in its position will make subtle changes in the amount of heat that is allowed to pass into the HRSG, and if the deadband is tight, that damper's actuator will be cycling a lot. I have seen a 4.6 MW turbine-based power plant brought to its knees owing to the failure of a $500 pneumatic actuator that could be traced to such a flow damper—repeatedly over a single year—because the root cause of the failure was never addressed. Such failures were a contributing factor to the overall failure of that power plant (as I describe in detail in a case study in chap. 7). Beware the design in which the failure of a $500 part will cause the entire $6 million power plant to be inoperable.

Size and costs. Combustion turbines are generally large in size, with electrical outputs in the range of 3 MW to more than 500 MW. For turbine applications at smaller sizes, microturbines can work well, and there are a wide range of sizes, heat rates, exhaust flows, and manufacturers to choose from.

Installed costs can be prohibitively high initially. Even with the minimal maintenance costs to balance that out over the life cycle of the equipment, it can still be $3,000/kW or more for smaller turbines to around $1,200/kW for larger units.

O&M. Most of the maintenance of combustion turbines involves borescope inspections of various ports along the combustion chamber and compressor stages, as well as the monitoring of bearing wear via direct measurement and analysis of the lube oil. There are two factors that are the primary determinants of the timing of the maintenance: the number of starts; and the number of run hours. A plot of operations (starts vs. run hours) serves as a predictive tool for when maintenance inspections are due. For example, figure 5–11 shows a *maintenance area* that is defined by drawing a curve to connect the two *inspections-due* points (e.g., 24,000 run hours and 1,200 starts) on each axis.

Later in life, turbines will need a more-detailed inspection than can be achieved via borescope. Keep in mind that these turbines are about the size of a school bus, but the outer casing is in two halves; these halves are disconnected from each other before the upper casing is lifted above the base, exposing the entire turbine's inner workings. Not only does this lifting and handling operation require great care, but a foreign material exclusion zone is also usually put in place to keep the potential of damage to the turbine blades at a minimum—no one wants a wrench dropped into the casing. Videos on the Internet can be found that show rigging failures, where the upper casing ends up crashing down onto the turbine below or the turbine shaft falling into the lower casing, with catastrophic results.

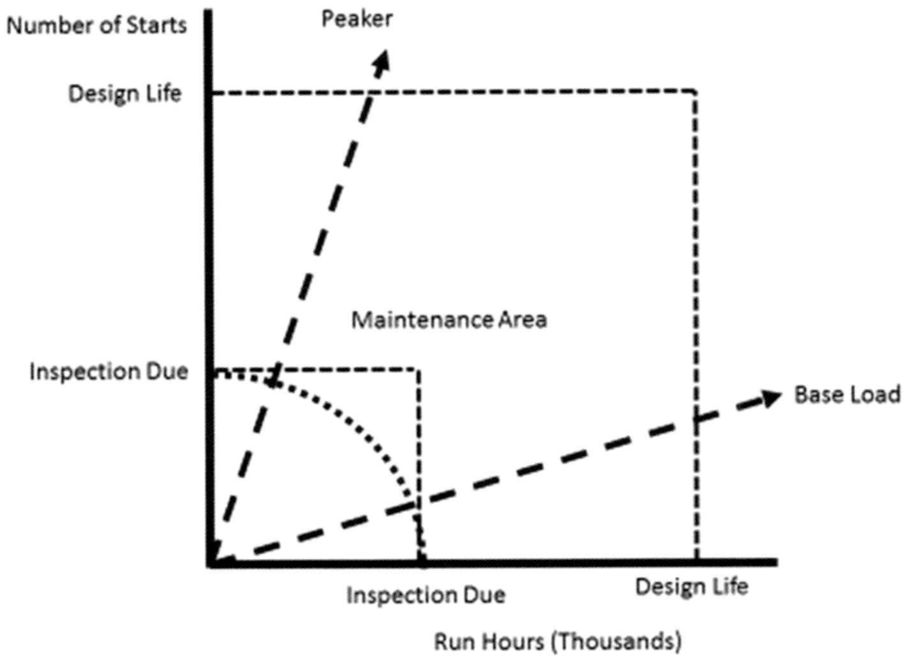

Fig. 5–11. Combustion turbine maintenance plan

Microturbines

Microturbines are small-scale versions of their larger cousins, the combustion turbine, and are available in sizes from 25 to 1,000 kW, although some of the larger units are actually just combinations of smaller units. By use of a modular approach, multiple microturbine units can be cascaded to build a total electrical generation capacity to meet many size requirements. Just like every other technology for a source of heat and power, there are advantages and disadvantages to sort out before employing a microturbine, as detailed in the next sections.

Advantages of microturbines

Fuel sources. Most microturbines are not especially picky about what they eat; that is, they can run on a wide range of fuel types and fuel qualities. For at least one manufacturer, the unit can burn everything from pure methane, to low-grade natural gas (75% methane), propane, liquefied petroleum gas (LPG), digester gas, sour natural gas, even diesel fuel and kerosene. The small size of the units, the ability to cascade them into large-capacity outputs, and their flexibility in fuel requirements make microturbines particularly attractive for agricultural facilities, from food processors to wineries. Siloxanes, however, are still a concern (as they are with any combustion-based unit) and must be removed from the fuel source as necessary.

Low maintenance. Most microturbines have very low maintenance requirements. This is because they usually do not have any kind of lubricating oils or other fluids that would require changing. Microturbines operate at very high speeds—on the order of 100,000 rpm—so any friction due to a bearing, no matter how well lubricated, would be too much. Plant designers get around this problem by using either an air bearing or a magnetic bearing, such that the rotating shaft has no physical contact with the unit's casing. The generator is used as a motor to get the turbine spinning prior to starting the combustion process, and the fins of the air bearing generate a cushion of air pressure to keep the shaft suspended and free to rotate.

The use of an air bearing is elegant but leads to a significant weakness: limited stop cycles. Every time the microturbine is turned off, the fins of the air bearing will ultimately fall owing to gravity as the air cushion dissipates, impacting the casing, and the fins can withstand that impact for only a limited number of times. For this reason, microturbines are better suited to a baseload situation where they can run continuously—which they can readily do for tens of thousands of run hours.

Another way manufacturers solve the friction problem is by using a series of magnets to keep the shaft suspended above the casing, floating by virtue of opposing magnetic fields. This is a bit trickier than using an air bearing, because small magnets do not have reliably strong field strength over time and the consequences of the failure of a magnetic bearing would be catastrophic. Otherwise, the maintenance items include the following:

- Change out the battery from the control system circuit board every two years
- Inspect the air filters every 4,000 run hours
- Replace the igniter once a year
- Replace the main battery every three years
- Inspect/overhaul the unit every 40,000 run hours

Low emissions. Most microturbines experience a very clean and complete burn of the fuel, producing about half a pound of NO_x per megawatt-hour (MWh) (or less), or about a quarter of a gram per horsepower-hour. This makes these units much more appealing to districts with very strict air-quality rules and permitting restrictions.

High efficiency. Microturbines, when paired with a heat recovery system, can have total efficiencies greater than 80%. Similar to the rotary engines used in some sports cars, the energy in the turbine is used to keep the rotor spinning in one direction; contrast this with the piston inside a reciprocating engine, which is continuously being forced to move, then stop, then move in the opposite direction, then stop again—always changing direction.

Unity power factor. Microturbines use their high speed of rotation to generate a very clean, high-frequency AC voltage that is then rectified into DC power, buffered by a set of batteries, and then put through an inverter to generate AC power at the desired frequency at a unity power factor. This makes the AC source indistinguishable from the AC sources used in solar PV power plants, and thus easier for the given utility to accept. This can also limit the surge ability of the AC source, as compared to conventional generation from a reciprocating engine. Another potential disadvantage with an inverter is the time rate of change of power output. That is, if there is a step change in electrical load, then the output of the inverter can increase only at a limited rate. During the time when the inverter output is increasing, until load is matched, there may be undervoltage-related problems with that transient.

Disadvantages of microturbines

High initial cost. Microturbines have a much higher initial cost than RICEs do—on the order of three to four times more expensive. This makes them vulnerable to short-term financial thinking, owing to the sticker shock associated with an expensive up-front cost for a new power plant. Life-cycle cost analysis, especially when you consider the near absence of any routine maintenance costs, can make these small turbines competitive with other power sources—but only if the financial parties can see beyond the horizon.

Few manufacturers. There are far more manufacturers of reciprocating engines than microturbines in this world. Furthermore, there are even fewer manufacturers who have a proven, good reputation, because these high-speed devices are a lot more difficult to get right than a conventional engine. This brings supply and demand into the financial equation, which keeps the prices of these units high. Supply and demand also needs to be a consideration when deciding what to include in your inventory of spare parts (and how many of each).

Lower thermal recovery when using a recuperator. One drawback to many (but not all) microturbine designs is that they use a *recuperator*, which is a heat exchanger that uses the turbine's exhaust to preheat the incoming air to the combustion chamber. Whereas this is a disadvantage in reciprocating engines, because it shifts operations toward premature detonation, it is a benefit to the turbine (which is continuously rotating in only one direction) since less energy is used getting to the ignition temperature. The drawback of using a recuperator is that there is now less net heat in the final exhaust of the unit to harvest by conventional cogeneration means; therefore, these systems are better suited for low-grade heat requirements (e.g., hot water, instead of steam).

Derating above 72°F. Most microturbines do not use an external turbocharger to boost the incoming air to the combustion chamber. Instead, they have a small built-in turbocharger—and therefore behave like a typical naturally aspirated device. Consequently, they are subject to derating as ambient air temperature increases, and many units will start derating at 72°F. For example, a 200 kW microturbine will start lowering its output (in a linear fashion) at 72°F to the point where it will be producing only about 150 kW at an air temperature of 120°F. This adds a serious constraint on the operations because, unless the units are in a power plant along the West Coast or in Alaska, chances are good that peak daytime temperatures are above 72°F (at least part of the year). In such warm places as central California, there are very few months in the year where the day's high temperature is below 72°F.

Unity power factor. Although also an advantage (see above), the unity power factor of the inverter output can be a disadvantage if the loads the unit will be powering are heavily inductive in nature, as is common in industrial facilities. Not all microturbines will work and coordinate well with electrical systems that normally operate at power factors of less than 0.8, and this needs to be taken into account from the beginning; otherwise the power plant may never function as desired.

Concentrating Solar Thermal

Parabolic trough thermal collector

Reflective material in the shape of a parabola allows for the direct focusing of solar radiation to a focal point. As illustrated in figure 5–12, incident solar energy—direct normal irradiance (DNI)—in the path of a direct beam (composed of parallel paths, DNI_x) can be reflected to the focal point, *F*, by striking any point along the parabola. By extending the shape of the parabola lengthwise, a reflector commonly described as a *parabolic trough* is made. By rotation of the trough to keep it positioned such that the direct path of solar radiation is maintained perpendicular to the open parabola, efficient collection of solar radiation can be maintained throughout the solar day.

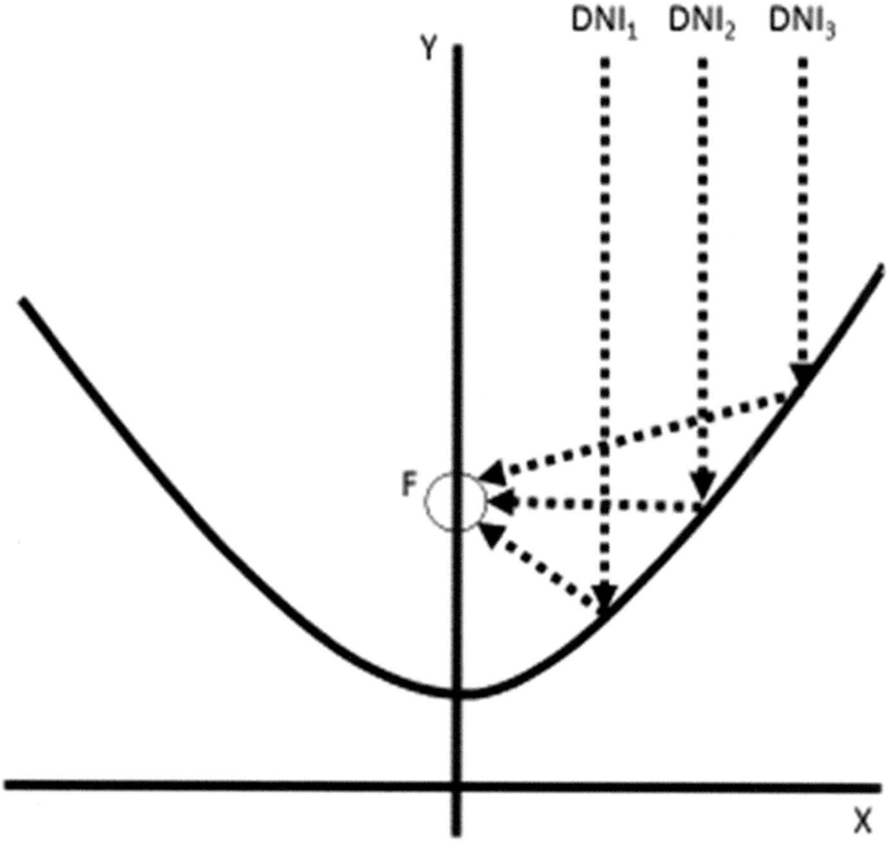

Fig. 5–12. Cross section of a parabolic trough solar collector

A service fluid, usually water or a thermal oil in commercial *concentrating solar thermal* (CST) parabolic trough power plants, flows through a pipe which is placed at the focal point of the parabola. The fluid in the pipe is then heated by the concentrated solar radiation and is then used to transfer that thermal energy to other systems such as boiling water for the production of steam. Substantial research has been performed by the National Renewable Energy Laboratory (NREL) over the past 50 years in optimizing the size, shape and other properties of the parabolic trough design for use in commercial CST power plants (fig. 5–12).

In the early 1980s, the Solar Electric Generating System (SEGS) power plants used parabolic trough reflectors in large-scale facilities in California's Mojave Desert. In the development of these power plants (nine in all) over the subsequent years, many lessons were learned and incorporated into each new plant's trough reflector design. As a result, these power plants experienced continuous improvement in the ability of the trough to withstand torsion from wind loads, reduction in reflector weight, and reduced costs.

In 2007, I met with Len P. Daniel, one of the engineers for the SEGS projects, credited by Discovery Channel as being "the builder of 90% of the world's solar power." We discussed Daniel's latest designs for the support structures of new parabolic trough mirror assemblies. His designs, as well as the commonly seen structures supporting smaller CST trough reflectors from Sopogy and Abengoa, consist of a framework of small beams supporting the delicate arch of the reflector. Key elements of each iteration of the design of the assembly were minimizing weight while simultaneously providing a rigid support structure for the glass mirrors to minimize torsion and thus minimize mirror damage by cracking during windy weather conditions.

While such solar thermal plants are commonly thought of as covering many acres of land, several of the manufacturers of trough reflectors also make small-scale versions of their equipment. Instead of being the size of a school bus, you can use reflector assemblies that are only six feet long and 12–18 inches wide at the mouth of the parabola. These smaller assemblies can be used together in small groups to provide a clean and easy source of low-grade heat that can be used to increase the thermal efficiency of any steam system.

Linear Fresnel solar thermal collector

The *linear Fresnel reflector* (fig. 5–13) uses multiple segments of mirrors, each aligned to reflect the incoming sunlight to the same focal point, to establish a similar concentration of solar radiation as the parabolic trough does. The mirror segments (lengths) can be individually rotated to maintain the light reflected from the mirror directed at the pipe running down the length of the solar field. Similar to the functional workings of the parabolic trough, there is a service fluid flowing through the pipe at the focal point of the mirrors, which collects heat from the concentrated thermal radiation from the sun. The heated fluid is then used in ways similar (or identical) to the facilities using parabolic trough mirrors.

A typical *concentrating solar power* (CSP) project is represented in figure 5–14. Direct solar radiation, or DNI, is the energy from the sun that reaches the earth. This sunlight is captured and concentrated by an optical system that uses a field of *heliostats* (mirror assemblies) that act together to focus the light, like a parabolic mirror, onto one place, the receiver. The receiver is the point of energy transfer from the sunlight to another medium.

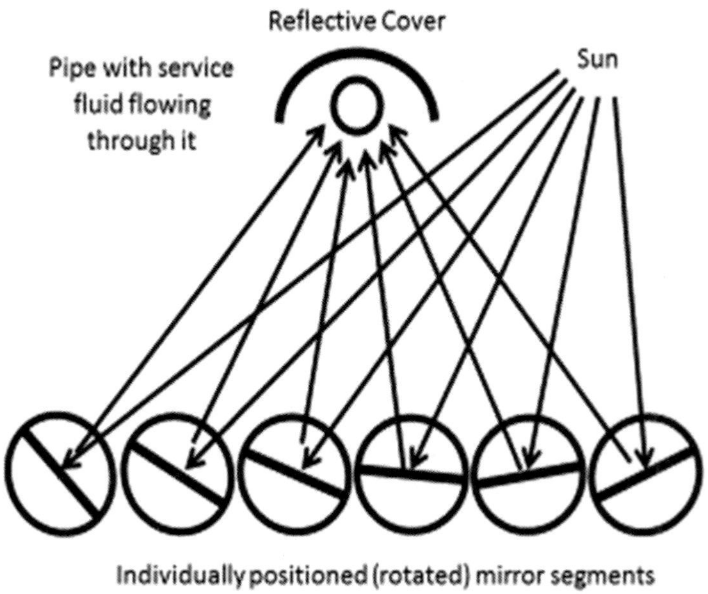

Fig. 5–13. Linear Fresnel collector

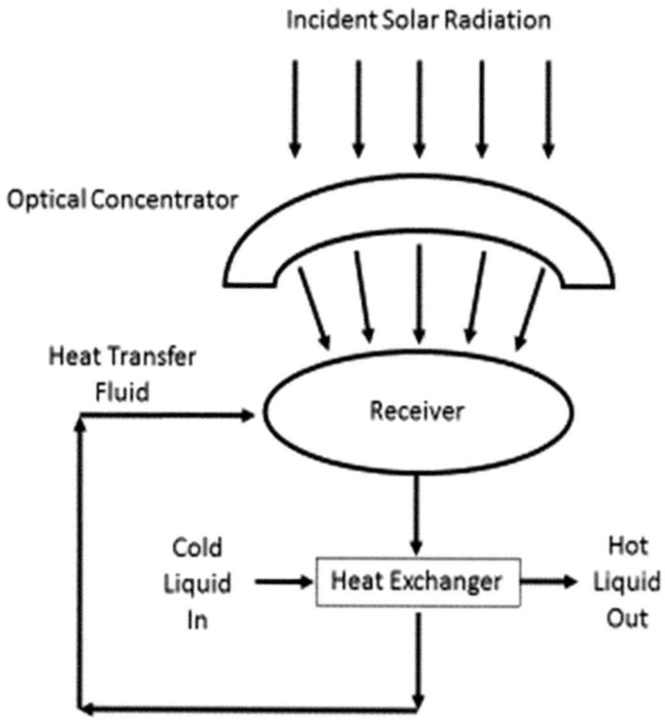

Fig. 5–14. CSP plant conceptual design

There are two main designs of the beam-down architecture for CSP: The first uses a secondary reflector (in Newtonian or Cassegrainian configurations), and the second uses a natural hillside to form a reflector by directly reflecting sunlight onto a collector at ground level. These are shown in figures 5–15 and 5–16, respectively.

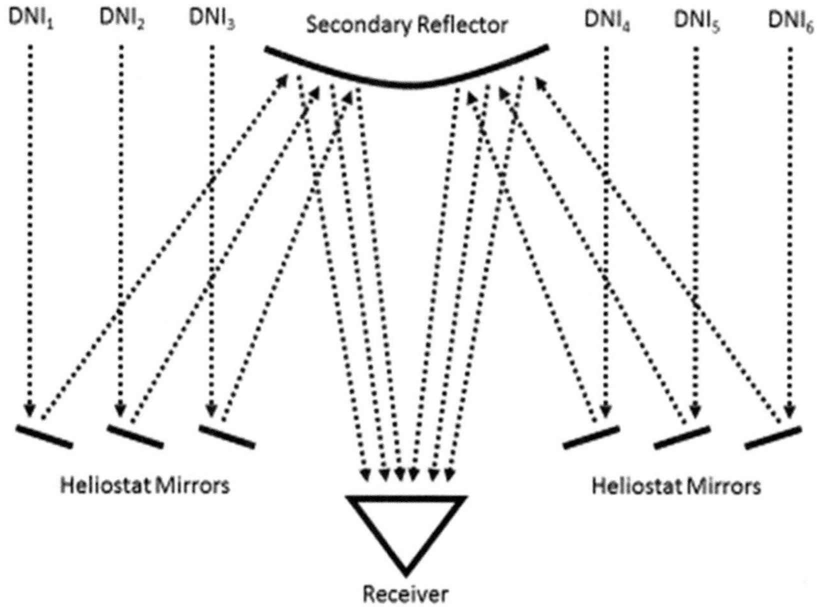

Fig. 5–15. Secondary reflector design

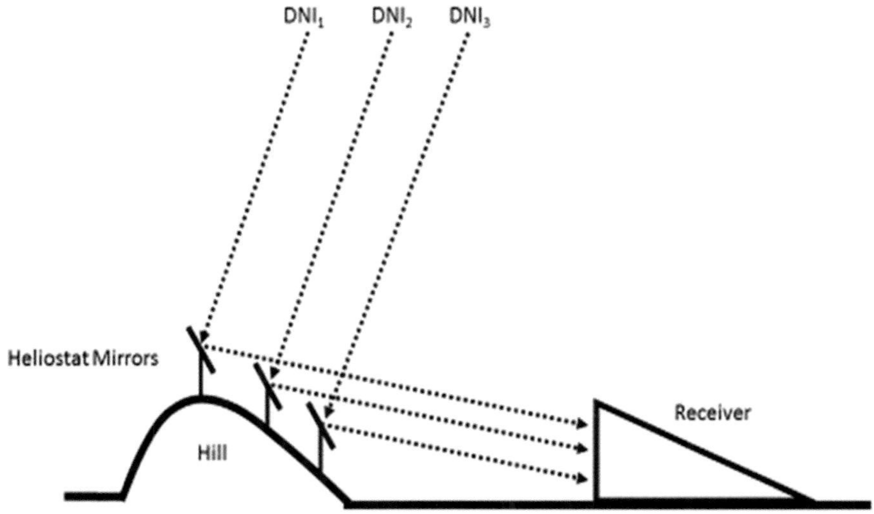

Fig. 5–16. Hillside reflector design

The hillside heliostat field has an advantage over the Newtonian design in that there are losses associated with having to reflect the sunlight twice, and the hillside design uses only one reflection. Conversely, the Newtonian design has an advantage over the hillside design in that the hillside design is limited to a smaller heliostat field, which cannot be more than 60°—unless multiple receivers are used—owing to the entry constraints of the receiver. Even with a centered Newtonian design, or an offset Newtonian design, multiple receivers (i.e., a compound parabolic concentrator [CPC]) may be necessary to collect all of the light from the secondary mirror (fig. 5–17).

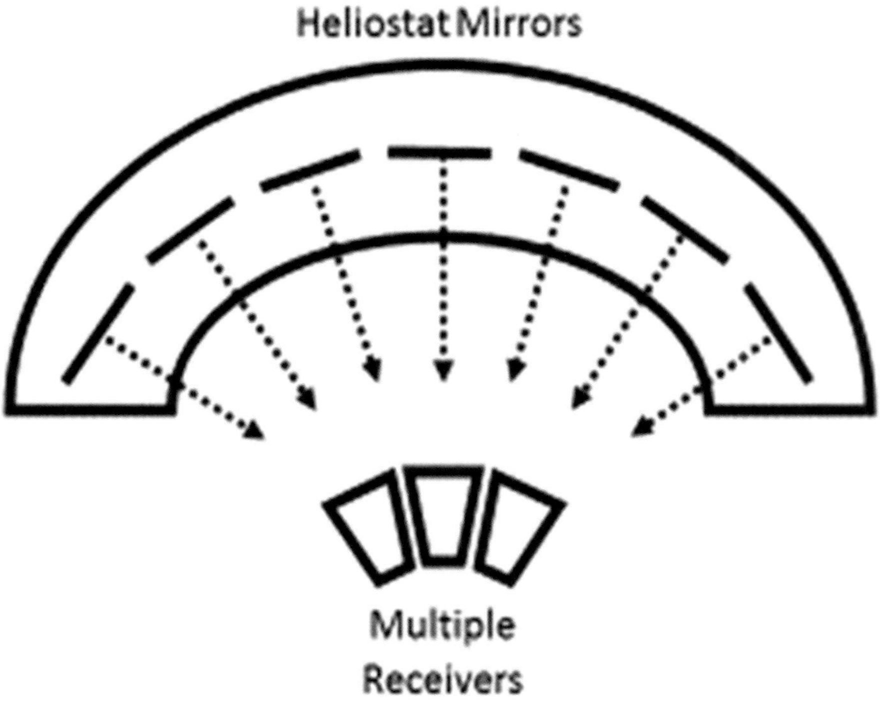

Fig. 5–17. Hillside heliostat

Old-school technology with new bells and whistles

The parabolic trough solar thermal collector design concept has been around for a very long time and has been used successfully around the planet. Depending on the location, the weather and incident solar radiation can be modeled to determine the amount of thermal energy available for harvest. If low-grade heat or intermittent thermal harvesting is needed, then small-scale systems can be used with single-pass or multipass designs; that is, the thermal fluid isn't necessarily restricted to a one-time pass through the collectors.

Thermal fluids, including organic Rankine cycle and molten salt

Although the use of glycol-treated water under pressure is still an option for the heat transfer fluid, most modern systems will use a lightweight oil with fluid characteristics very similar to water but stable up to about 740°F. For the largest solar collectors, including heliostat arrays, one option is the use of *molten solar salt,* a mixture of sodium and potassium nitrates. The molten salt system moves the heat from the receiver to a thermal storage system, which is also supplied heat from a fossil fuel source, to a heat engine. The heat engine is typically a heat exchanger, where the heat from the molten salt is transferred to water to generate high-pressure steam, and the steam is then used to generate electricity via a turbine.

For lower-temperature systems, especially small-scale CST facilities, organic fluids can be a good choice. These are usually hydrocarbon-based working fluids (e.g., refrigerants) that have boiling points lower than water. Therefore, they can use small amounts of heat to reach a boiling point and then the vapor can be used to drive a small turbine, just as one would use steam in a higher-temperature system.

Costs

The capital costs of installing a CST power plant will vary with the size and complexity of the facility, but a rule-of-thumb amount is \$5,000–\$6,000/kW (installed). Similarly, the costs of operating and maintaining the facility will also vary widely, especially when you include weather-related mirror replacements, and could average \$60–\$70 per kilowatt-year.

Locations and weather limits

There is a reason why there are no CST power plants within 25° latitude of the equator, even though that area gets the most intense sunshine. The reason why no one does so is obvious when you take a look at the weather in those tropical regions and see the cloud cover and daily rainstorms. Deserts make great places to install acres of CST assets, but if your industrial or commercial facility isn't out in the middle of nowhere, this doesn't help your situation. Still, there are a lot of places that get enough sunshine—and for enough hours a month—to make the installation of some form of CSP advantageous.

O&M

The operational side of CSP installations is in two main parts. The first part comprises the system of gears and/or hydraulics positioning all of the heliostats (as well as the control system, since each mirror will be individually positioned). The heliostat field can consist of thousands of individual mirror elements, so a computer is used to position each element. The second part is the thermal receiver taking in the concentrated sunlight and the associated heat recovery system, which has control over the hot fluid's flow. The control of the heat transfer, both into the receiver and to the process fluid, is part of the job description for the power plant operator. The preventative and corrective maintenance on the pumps, valves, and other fluid system components is complicated by the high temperatures of the fluids. For the molten salt to be kept in a liquid state, its minimum temperature must be maintained. Changing out pumps and valves requires additional safety precautions, as well as fluid management. For the heliostat field, the maintenance requirements will be proportional to the complexity of the mirror positioning systems. There will be differences depending on whether the drive systems are electric (via gears), hydraulic, or a combination thereof.

Hybridization

CST power systems can work well in a hybrid arrangement with more conventional power plants, especially where the solar bits can carry the bulk of the energy demands throughout the day. This is enhanced with the employment of on-site energy storage, such that solar power can supplement the conventional power plant for at least a few hours after sunset.

Passive Solar Thermal—for Low-Grade Heat Recovery

Passive solar thermal describes methods to gain thermal energy from the sun that do not require any tracking device. For example, painting a large water storage tank black will allow it to gain more heat than if it were painted white. Another example I have seen is the use of very large radiators working in reverse—that is, instead of circulating hot water inside to transfer that heat to the atmosphere, relatively cold water is circulated through the tubes to gain heat from the atmosphere.

The same idea is used by homeowners around the planet in the form of a small tank and flat heat exchanger mounted on the rooftop to supplement the domestic hot water supply. One hazard of these systems, if there is any chance of leakage across the heat exchanger at all, is that unless the water is chemically treated or heated to over 140°F, there is a very real chance that legionellae, the bacteria causing Legionnaires' disease, can thrive in it.

Photovoltaic

Field sizing—area versus kilowatt output

Plenty of books have been written on the subject of photovoltaic power, so here we focus on basic considerations if you're going to use a PV system. As a general rule of thumb, I use a ratio of seven acres of land to yield 1 MW of AC power. PV panels come in a wide variety of shapes, sizes and outputs, with innumerable mounting options. The PV panels can be strung together in combinations of series and parallel units to get whatever the power output desired. There are a variety of federal, state, and local ordinances that dictate what *high voltage* means, as well as what is required if you exceed that amount, for both the DC and AC sides. In general, there are two ways of combining the PV panels:

- Multiple, low DC voltage strings coupled to multiple, small inverters that share a common AC output bus
- Longer strings to generate high DC voltages (600 V DC) using fewer, but larger, inverters

Risk management must be a part of the planning, because the economics of an outage of one part of the PV and inverter array must be accounted for. The failure of one large inverter will affect the total output of the PV array far more than the loss of a single, small inverter. The net costs to the overall PV system—including the maintenance requirements, spare-parts inventory, and availability of maintenance and repair personnel—go beyond just the equipment selections. Beware configurations that allow a single-point safe condition, where one small equipment failure can bring down large parts of the array.

One of the big pluses to using a PV system is that the panels can go pretty much anywhere. That said, they should not go just anywhere, and I see examples every day (in my neighborhood) where someone has put PV panels up on rooftops facing the north—and often under a tree. Industrial and commercial facilities usually have a fair amount of rooftop space for PV panels, as well as a parking lot for the employees that could be covered to place panels on top. Use a good solar software and/or do a study on a year's worth of solar positioning—for example, checking for shadowing by trees and buildings, as well as by other panels in the array. I have worked with customers whose buildings were pretty old, and rooftop structural integrity was not guaranteed, especially for considerations of wind loads.

Fixed versus single-axis or dual-axis tracking

There are many ways to mount PV panels so they can do their job. As the earth rotates around the sun, the azimuth of the midday sun changes, and PV panels need to be positioned directly facing the sun to get full output.

The least expensive panel mounting is the *fixed* mount, where the PV panel is stationary at some angle to the horizontal. This angle is usually selected by a computer program for optimum output, the dictations of the placement location (e.g., a building's sloped rooftop), or convenience. Alternatively, the output of the PV array can be increased by using equipment that changes the position of the panels to track the sun over the course of a day; these can be either single-axis or dual-axis systems. Mechanical tracking systems increase the cost of the PV array and add a maintenance cost to the O&M budget. A simple single-axis tracking system can give a positive return on investment by increasing the net output of the array. Adding a tracking system for the second axis is less cost-effective. Unfortunately, for many of the PV systems I've dealt with, the added costs to the project, both initially and through the equipment life cycle, are not offset by the increased revenue brought in from the gains in the system's output. Ultimately, cost/benefit analysis must be performed on every case, so you should not be deterred from using PV systems where it makes financial sense.

Biomass

Biomass, as an energy source, is just what it sounds like: a mass of biological material that can be used, directly or indirectly, as a source of fuel for a chemical transfer of energy (i.e., burning). Some materials can be burned directly, such as wood or cow manure, and others are better suited to gasification, whereby they are treated to produce a flammable gas.

Right now, in California, there's an abundance of dead trees (mostly pine) owing to extended drought conditions and the infestation of a bark beetle. The logistics of getting access to these trees, harvesting them, and bringing them to a biomass power plant can be tough, since the Sierra Nevada Mountains are pretty rugged. I have been to several biomass power plants, and the wood-based ones are a blend of art and science. The science part is fairly easy and entails designing and constructing the incineration chambers and boilers, among other features. The art comes in the blending of the fuels. Different sources of wood and wood waste (e.g., old pallets) have different heat energies when burned, and the power plants employ people to identify, sort, and mix the various feedstocks to get the desired heat from the burners. For example, if you've ever used a fireplace, you may have noticed that oak burns differently from pine or almond wood. In addition, when waste wood is used, a magnetic sorter is needed to pull out metal scraps such as nails and staples prior to combustion, or else the combustion chamber will need more frequent periodic cleaning to remove the collection of noncombustible debris that will accumulate there.

Another burnable material is cow manure. In 2006, I had to come up with an O&M plan for a 50 MW hybrid solar/biomass power plant that used manure from the 430,000 head of beef cattle that were within a 50-mile radius of the site. As with most "unclean" fuels, there are problems associated with manure, including acids that are produced in the combustion process that will corrode the heat transfer surfaces.

Logistics, too, constitutes a large consideration in the biomass power plant operations, in that a steady supply of fuel is required in order to maintain a continuity of power output from the facility. This usually means a constant stream of large trucks hauling payloads of the fuel, whatever it is, and this can be a problem for the neighborhood of the power plant; imagine a new truck arriving every six minutes, all day, every day, to deliver an abundant supply of the chosen fuel. This also adds to the O&M costs of such a biomass power plant, not just for the given fuel but also for the trucks, their drivers, vehicle upkeep, and so forth.

On a small scale, more suited to an on-site cogeneration power plant, biomass can certainly have a place, often with unanticipated benefits. A facility in California that processed olives used to accumulate olive pits that had to be trucked off-site as a waste product; however, olive pits can burn. The olive facility worked with a local energy services company to design and construct a combustion and heat recovery process, converting those waste olive pits into steam. This not only saved money on the existing thermal plant, since less natural gas was needed for their existing boilers, but also eliminated the expenses associated with removing the pits from the facility and paying a disposal fee. Other food processors can do similar things, although caution needs to be used as unanticipated hazards may exist; for example, burning peach pits can release cyanide gas. Many people, when they discuss biomass as a fuel supply, are really talking about biogas, which is next on the list.

Biogas

There are several ways to generate a flammable gas from biological sources, and these are collectively referred to as *biogas,* which is a combination of methane, CO_2, and other constituents, with variations in the relative proportions of each (depending on the source material). The manure of dairy cows, or any other liquid manure, can be directed into large holding tanks where it can be allowed to ferment (along with other farming wastes), and the flammable gases produced can be extracted for use. Enclosed landfills can also produce actionable amounts of flammable gas, not just aerobic or anaerobic digesters. In many cases, the gases produced must be cleaned up before they can be used (e.g., removing hydrogen sulfide), and this can add considerable expense to the project.

Anaerobic digesters are those where the processes occur without oxygen being added to the system, but rather the oxygen is obtained from the materials being digested. Almost any biodegradable material can be used as a feedstock, but the more putrescible it is, the better. People have more experience with aerobic digesters in day-to-day life—the main example being a compost bin. In industrial use, organic material is allowed to decompose in an aerobic digester, and the gases produced are again extracted and cleaned for subsequent use. This process is a little easier, in that excess oxygen in the digester is not a problem, and there's no real limit as to what biological material can be used as a feedstock. Many food processors are now looking into this as an alternative to paying for the removal of organic waste from their facilities. The trick is to understand the nature of the gases that are generated by the digester and what kind (and amount) of cleanup is required before using those gases in a combustion process. Some equipment, such as microturbines, will burn just about anything, but a reciprocating engine will not be able to burn most biogases without their being *sweetened*, by adding natural gas.

The need to blend the fuels going into the engine, adding to (or reducing) the amount of natural gas as the biogas supply varies, is an area of engine performance getting a lot of attention from the OEMs. Part of the problem is the control system's feedback loop—in particular, there is typically no fuel quality metering system on the input side of things, so what indicates the need for more or less natural gas in the mix? The answer lies in the response(s) of the engine to the fuel mix going into it, which will change as the supply changes.

The control system's feedback loop comes in because the system must recognize that the engine would prefer a different blend of the fuels and respond quickly. The time delay—from the threshold when this need is sensed through when the signal is sent to reposition the fuel blending valves and those valves actually moving, to when the improved fuel mix gets to the cylinders—must be shorter than the defined period for the antiknock or detonation sensors to cause the engine to shut down. This kind of adjust-on-the-fly system for fuel blending valves takes a lot of work to get right. In practical terms, it will never be perfect, but advances have been made in this area, and we should expect more of them in the near future. The presence of available biogas (however variable in quality and quantity) at more and more sites around the country is driving OEMs to get their fuel blending systems tuned so they are not the last ones to market their solutions to the power industry. Still, the first OEMs to respond to this opportunity with a marketable product may not have the equipment perfected yet. By analogy, the computer software industry is not the only one to get a product to market first, then fix all the bugs in the system, rather than take the time to get it right the first time.

Another factor in the supply of actionable biogas is that the supply chain of feedstock, as organic material, will probably contain some variability. The agricultural processing industries, whose waste streams seem ideal for this kind of application, will have varying feedstock based on crop seasons, weather, and other factors. For the power plant developer considering using biogas as part of the solution to the on-site power generation portfolio, the variable nature of the availability of the fuel must be carefully addressed. One way to handle this variability is to use the biogas in a boiler or via a burner in a hot water heater. Doing so allows the time-sensitive balancing of the biogas and alternate fuel (sweetener) to be managed by a much more mundane control system than on a RICE.

One major source of trouble in biogas is the presence of siloxanes, especially in a wastewater treatment facility. Siloxanes are compounds of silicon and oxygen that are found in everything from medical products to shampoo, and these easily find their way into wastewater streams. Wastewater treatment plants can use the sludge created in their processes as a feedstock for an anaerobic digester; however, the temperature in such digesters can get extremely high, and once it reaches about 140°F, the siloxanes can volatize. If the desired flammable gas is burned along with the siloxane vapors, the siloxanes will basically become miniature particles of sand that then melt onto the internal parts of the reciprocating engine—effectively coating them with glass—and this will result in the demise of that engine in a very short period of time (minutes).

A symbiotic relationship can be shared between a wastewater treatment plant and an on-site cogeneration power plant. Specifically, the power plant can be fueled by the wastewater plant, and the heat from the cogeneration power plant aids in the wastewater treatment processes, with the electricity that is produced as a nice by-product. The problem is that the gas cleanup is nonnegotiable and must be done in a way that ensures that the siloxane vapors cannot reach the engine, which translates into an expensive set of equipment (that is also single-point safe). I have participated in a development team for such a power plant, and the gas cleanup equipment ended up being over 50% of the budget for the whole project, making it prohibitively expensive. That's why it is not uncommon to see wastewater treatment plants simply flaring their gas.

Geothermal

There are geothermal hot spots all over the planet, and under the right conditions there is an abundance of heat ready to be harvested. Unfortunately, most of these hot spots are in remote areas and thus are not very practical for a power plant located on-site at an industrial facility. Utility-scale geothermal power plants are expensive to build, and a range of dangerous complications can arise during the construction phase; however, they are relatively inexpensive to operate once built and online.

If you're lucky enough to have a geologic heat source near the site of your power plant, such as a hot spring, then find a way to use that heat. Even if it is low-grade heat (i.e., not enough to generate steam), it can still be used to preheat the feedwater supply to an HRSG. Another variation of the geothermal mechanism is the ground-source heat pump (of which there are several varieties). The temperature of the soil at the surface of the ground, plotted over the course of a year, will have a roughly sinusoidal shape. As you go deeper underground, the temperature plot is also sinusoidal, but it gets further and further out of phase (lagging) with the surface temperature. The temperature of the soil that's about six feet underground can peak a full month after the maximum temperature is reached on the surface. By the time you get about 15 feet underground, the delay is around two months. Furthermore, the amplitude of the temperature wave decreases as depth increases, to the point that at depths of 30 feet and below, the soil temperature effectively remains constant throughout the course of the year. Therefore, a heat exchanger buried to this depth could be used as a heat sink or very-low-grade heat source (for organic Rankine cycles).

Wind

Use where practical

I have lived places where it seemed as though the wind never stopped (namely, Chicago), as well as places where the wind was noticeable only at certain times of the day (Hawaii). When it comes to employing wind power, I recommend considering it where and when it makes sense. Industrial and commercial facilities do not typically have rooftops that are high enough off the ground to be at an elevation with more air movement than at ground level, but it does happen. Placing a few wind turbines on the roof of a building can certainly provide supplemental power to the overall site, and can even provide power when the sun isn't shining. Overspeed protections are needed, to protect the assets under thunderstorm conditions, and this is something that might get overlooked at first because it is not a typical power plant problem.

Highly variable, therefore storage is needed

If you do use some wind power as part of a hybrid on-site power plant, I highly recommend including an energy storage component (usually batteries). This can help buffer the intermittent nature of the resource while adding flexibility to the overall design of the power plant.

Other Sources

Hydroelectric

Although most people think of hydroelectric power as being associated with enormous dams and rivers or reservoirs, it doesn't have to be limited to these geographic features. If the site of the facility has any significant change in elevation across the property, then on-site hydroelectric power should be considered. Even if there is no immediate source of water available, most buildings have rainwater downspouts, and these can be used as a flowing column of water. Microhydro turbines are a reality and have been used by rural homes for a long time as a source of low-voltage DC power. Even though these microhydro turbines will not generate industrial quantities of power, when it comes to off-grid situations, every little bit helps. Combining solar PV, small-scale wind turbines, and microhydroelectric systems with a battery allows the implementation of a DC-based power distribution system. Certain lighting, ventilation fans, and other equipment can be powered by such a low-voltage, DC-based power system.

Fuel cell

The idea of the fuel cell has been around since the 1830s, but only recently has the concept and technology been pursued on a large scale. While the various technologies use different chemistries, the basic idea is the same. Hydrogen is ionized, and the electrons are put to use as a DC power source before recombining with the hydrogen and some oxygen to make water. Also, since hydrogen is the target fuel, the use of hydrocarbon-based fuels (e.g., natural gas) makes for a convenient and steady fuel supply. Fuel cells are, though, sensitive to the presence of sulfur, so any traces of the contaminant must be removed by a gas cleanup skid prior to use, and this will naturally increase the cost of the unit. The differences among fuel cells relate to the chemistry of choice, which can also affect the operations and maintenance of the fuel cell, including startup/warm-up times and on-off cycle limitations.

Also highly variable are the operating temperatures—and, therefore, the potential thermal energy harvestable. Operating temperature range from near ambient for the PEM (proton exchange membrane) fuel cell to almost 2,000°F for the SO (solid oxide) fuel cell. The units that operate at the higher temperatures, namely the SO and the MC (molten carbonate) fuel cells, have very long warm-up and cooldown times, which makes them better suited to baseload than cyclic operation.

One general advantage to using fuel cells is that they can operate at reasonable efficiencies all the way down to about 25% of their rated capacity, unlike combustion-driven equipment. Other considerations that can make the use of fuel cells attractive are as follows:

- They produce essentially zero emissions, so they are a good option for highly restrictive air pollution control districts.
- They are fairly quiet in their operation, which makes them a good candidate for use in places like hospitals and universities.
- They can digest a wide range of fuels, not just natural gas (although a cleanup skid to reduce or eliminate any contaminates will increase the initial costs of the project, as well as introduce a single-point safe item).

On the negative side, beyond the initial high costs of fuel cells, the internal components of the units will have to be replaced every five to seven years. Together with the minimal (if any) heat recovery from fuel cells, this can offset any benefit of having low emissions as far as application in a cogeneration power plant goes. Research continues, however, on new chemistries and on improvements to existing, known elements that work, so the future remains positive for the impact and application of fuel cells in distributed generation projects.

Coal

I have been to several coal-fired power plants, and although none of them were part of a cogeneration system, it is certainly possible to use coal as a supplemental source of heat. In some parts of this country, coal is plentiful and readily available for use, bringing the costs associated with fuel logistics down. When used in an on-site boiler, coal can augment the thermal output of a cogeneration power plant and thus may have a place in the plant design.

Personally, I believe the negatives associated with coal outweigh the advantages associated with its supply logistics. Among these disadvantages are the ever-present black coal dust that will get into everything; the trace amounts of mercury, uranium, and thorium and the associated environmental hazards; the required cleanup of the combustion exhaust; and the disposal of the end waste product, the ash.

The term *clean coal* is an oxymoron that has been invoked for decades. The distributed uranium, thorium, and mercury in coal are not exactly "clean" components, let alone the particulate exhaust. The mercury gets distributed to everyone downwind, and the uranium and thorium concentrate in the ash, making the radiation emissions from coal-fired power plants higher than those from nuclear power plants.

A friend of mine, the chief executive officer of Sidel Systems, has patented a couple technologies that can help the situation. One is a biological amine-based system to absorb the CO_2 from the exhaust and form a marketable by-product. The other removes the particulate from the exhaust. When these two technologies are combined, suddenly coal becomes markedly cleaner. There may yet be ways to make the burning of coal competitively clean with other technologies, if given enough market penetration.

Natural gas

Pipeline-quality natural gas is available in most industrial and commercial areas around the country, and there are utility companies that maintain high-volume, high-pressure pipelines to distribute that gas to population centers. Nevertheless, in populated areas where there is high demand for natural gas for residential uses, these same gas utility companies can issue curtailments to their industrial and commercial customers. This practice used to be limited to especially cold days during the winter, but now, even if the ambient temperature drops to 40°F, some industrial and commercial facilities find themselves without a reliable supply of natural gas.

The increasing curtailment of natural gas has been favorable for those in the liquefied natural gas (LNG) industry, because those facilities facing interruptions in their supply of pipeline natural gas are increasingly installing tanks of LNG at their sites that can be refilled by truck deliveries. With enough tanks of LNG on-site, the random curtailment of the regular natural gas supply can be overcome, even for long periods of time. However, LNG is a cryogenic liquid and must be carefully *regasified*—that is, brought from –260°F up to a more reasonable temperature and into gaseous state. The heat exchangers and high-pressure tanks associated with this regasification process, along with the LNG storage tanks, are all part of the financial decision whether to install LNG as a backup fuel source.

As a fuel, natural gas used to depend heavily on the Gulf Coast area as the main source for the rest of the country. The result of this can be seen in the effect Hurricane Katrina had on the price of natural gas—and the subsequent ripple effects throughout the energy sector (and others). Thanks to the widespread use of hydraulic fracturing (fracking) around the country, natural gas production is no longer tied to any one particular geographic area. Natural disasters can happen anywhere, but now the impact on the overall supply of natural gas to the nation will be minimized if an isolated area is affected alone. This not only makes the supply of natural gas more assured, it also mitigates any spike in the price of the gas based on simple supply and demand economics.

Stirling engines

Stirling engines have been around a long time and have shown some promise in on-site power generation with solar and biomass/biogas applications, but there's been no real traction yet. As opposed to using internal combustion, the Stirling engine is an external combustion device. That is, the source of heat to the engine is applied from the outside of the cylinders, as well as the application of a source of cooling. Inside the cylinders, which are sealed, a gas (usually hydrogen or helium) is alternately compressed, heated, expanded, and cooled, and then the cycle repeats. As the cylinders move, they convert linear motion to rotational motion via a camshaft, much the way a RICE operates.

The advantage to using an external heat source is that any source of heat will do, so most of the research into making Stirling engines viable was in the biofuels and CST. The cleanup requirements of biofuels prior to use in a RICE are not required here; thus, a simpler and more complete burn of the fuel can take place, and these come at a lower cost to build and operate. On the solar side, because of size restrictions, the Stirling engines can be deployed in parabolic dishes, sitting at the focal point. In prototype tests, these dishes have been composed of individual mirror segments that formed a dish (resembling large satellite arrays) about 30 feet in diameter. These prototype units were small, electrically, at only 25 kW of output, but modular enough to be combined and cascaded into larger arrays.

As of this writing, no companies yet manufacture Stirling engines for stationary power systems on a large scale. However, several companies are researching the application of these engines in systems that can be cost-comparable to PV systems. Even with their advantages regarding fuels, Stirling engines will not be able to compete with existing proven technologies until they become more financially viable.

Energy Storage

Unless your on-site cogeneration power plant project is especially fortunate, not all of the electrical and thermal energy that is produced will be used at the time it is made, so consideration should be given to employing some form of energy storage. Over the past decade or so, there has been a tremendous amount of research and development focused on the challenges of energy storage. Breakthrough technologies in a variety of systems—from compressed air to new battery chemistries to molten salt are now available—and these can be divided into electrical, mechanical, and thermal systems.

A tremendous amount of research has recently been done on improving batteries, and this is expected to continue to be a focus of interest. New chemistries, novel configurations, and increased capacities are being announced every day, but these do not all come without risks. In any case, solid-state batteries are going to be more and more a part of our everyday lives—but are they a good fit for the demands placed on a battery system by a hybrid cogeneration power plant? Not necessarily. There are a lot of choices for your battery needs, and each will have advantages and disadvantages. Batteries based on good old lead-acid chemistries are reasonable choices for the demands of a power plant support system; however, they do take up space, and there's always a concern about hydrogen venting during a battery charge cycle. There's also plenty to choose from among the nickel-cadmium and sodium-sulfur chemistries. In any case, for power plant applications, NREL has a downloadable tool (posted on their website at the time of this writing) called SAM (System Advisor Model); under the PV section using this tool, you can model the battery options by entering or changing operating parameters to see the results from using a variety of lithium-ion and lead-acid battery chemistries.

Electrical energy storage

Inventive people are searching for a way to store AC power, but until they succeed, electrical energy storage means DC storage. Batteries and battery charging systems have been around for a very long time, so that part of the idea isn't exactly new. What is new is the wide variety of chemistries that are now available; these choices also allow new physical packaging and ways of doing things than just a lead-acid cell.

With these new technologies, beware of claims from battery manufacturers that seem too good to be true—for example, that their product has a high energy density, delivers thousands of charge/discharge cycles at high depth-of-discharge amounts, and has a 20-year life span. Similarly, just because a battery uses a nickel-metal hydride (Ni-MH) chemistry does not mean it will have a high energy density (unless it uses a prismatic architecture). Caveat emptor! In general, the battery chemistries most often used for power plant applications are lead-acid, nickel-cadmium (Ni-Cd), Ni-MH, and vanadium redox, each having advantages and disadvantages.

Lead-acid batteries are well known and are manufactured in a wide array of sizes, shapes, and capacities. Early versions required some attention, having to maintain the level of electrolyte in each cell and taking care to vent the hydrogen that was given off during the charging cycle. Modern, sealed versions are now essentially maintenance free. The lead-acid battery does not suffer the memory effect that plagues Ni-Cd batteries (see below), but whereas the Ni-Cd units require at least a periodic full discharge, lead-acid batteries won't last long if deep-cycled. The depth of discharge has a huge impact on the life span (i.e., number of charge/discharge cycles), so careful planning and modeling of how lead-acid batteries would actually be used in the power plant is crucial to the system performing as desired.

The rate of charge to discharge and the ambient temperature also play big roles in how long such battery systems can remain in use. Long-duration charges and discharges and moderate ambient temperatures are preferred. If the size of the footprint is not a concern, then a nice air-conditioned building the size of a four-car garage can get you plenty of MW of DC power at 75°F for about 10 years (if treated gently); That same size system (operated just as tenderly) exposed to 110°F weather would last maybe 1 year.

Another potential limit is the charging voltage. Every state has a threshold of what constitutes high voltage, and on the DC side of things that is often 600 V DC. When a 12 V DC lead-acid battery is charged, the charging voltage is typically 14.5–15 V DC, which would limit the number of cells wired in series to 40 units.

Ni-Cd batteries enjoyed almost ubiquitous use in small portable electronic devices before the age of lithium. As such, most people experienced one of the Ni-Cd battery's operational points of frustration—namely *memory*. That is, if a Ni-Cd battery was discharged from a fully charged state down to about 80% charged before being recharged back to 100%, over and over again, a problem develops. Crystals end up forming on the cell's plates, such that continued discharge below that 80% mark would not happen at the specified output of the battery. Effectively, the nature of the chemistry would ensure that the range from 80% to 100% would become the battery's new "full" range of charge, as it would "remember" only that band of operation. Therefore, although deep discharges are a life-cycle limiter in a lead-acid battery, they are absolutely necessary in a Ni-Cd battery.

Ni-MH batteries provide some improvements over Ni-Cd units, but at a much higher cost. Ni-MH units have significantly higher energy densities than do Ni-Cd cells and exhibit less of a memory effect. In either nickel-based chemistry, the batteries will need periodic deep cycling and a more elaborate charge control system than the lead-acid units do. Ni-MH batteries also come at a high price; in other words, energy density is a premium product.

Another option that has been increasingly popular is the *flow battery*, wherein the rechargeable electrolyte is dissolved into liquids that are held in tanks, flowing past each other via a membrane. These are exemplified by vanadium redox batteries, which came on the scene around 1990 and have the unique feature of storing the electrolyte in tanks, with the exchange of charge happening inside a separate unit. The size of the battery is limited only by the size of the tanks storing the two fluids and has extremely high charge/discharge cycles per lifetime. If the size of the footprint is not a concern, then flow batteries can be an economical choice—especially given their long life spans. In recent years, many research and development companies have sprung up to work on flow batteries, experimenting with different chemistries.

Flow battery systems are inherently safer relative to other battery options, because they are less sensitive to changes in ambient temperature than conventional battery systems, and all one has to do to shut down the reaction is stop the flow of either fluid. In addition, compared to lithium-based batteries, the lower energy density of the flow battery makes for a much easier O&M, especially when you consider the protocols associated with responding to incidents like an on-site fire. With conventional battery systems, it's almost impossible to separate individual batteries to split the short-circuit voltage supply if there is a battery fire, instead having to clear the building while the fire burns itself out. Here, the size of the battery is based on the chemistry used and the sizes of the tanks holding the electrolyte. Nevertheless, do not overlook the responsibility of containing any spills that might happen if a tank fails—and the "fun" of the hazmat cleanup efforts that will need to happen.

Lithium-ion batteries seem ever to be increasing in popularity, but that doesn't mean they are a good choice for an on-site cogeneration power plant. Lithium-based batteries have the promise of very high energy densities, or having a lot of energy in a small package. However, most lithium-based battery chemistries are problematically finicky. For example, many of these batteries do not do well if overcharged or discharged excessively. If not handled with care, batteries using this chemistry have the potential to catch fire spontaneously—or "reconfigure themselves," as one conference speaker put it.

Certainly, lithium-based batteries can have a place in a power plant, but a careful examination of the various chemistry choices that are available (and proven) should be performed before succumbing to hype from OEMs. Lithium-based batteries have more than just the higher costs, as a premium for higher power density, but those benefits are wasted in an on-site power plant (at least in the cases I have analyzed). For one, space is not usually at a premium at an industrial or commercial facility, so there's plenty of room for less expensive technology. Second, as already mentioned, lithium-based battery systems are finicky about being overcharged or discharged too far, as well as being particular about what the environment is—in other words, ambient temperature matters. If you want your stack of lithium-ion batteries to last a long time, you'll limit the temperature to about 60°F and limit the charge and discharge rates to one or two times the capacity of the stack. The last thing you want is to shoehorn a new lithium-ion battery into your power plant only to have it "reconfigure itself" (i.e., catch fire) once factors deviate from the battery's comfort zone.

Because lithium-ion batteries have about twice the energy density of the Ni-Cd batteries, they have replaced Ni-Cd batteries as the power cell in our ubiquitous portable devices. However, that energy density comes at a price, and beyond the economics of it all, one of the biggest costs is instability. Also, even though lithium-ion batteries do not exhibit the kind of memory issues associated with the Ni-Cd chemistry, there is some loss of capacity over time, so the life-cycle time may not be as long as advertised.

In the planning and evaluation of an on-site cogeneration power plant that will include some kind of electrical energy storage component, it can be difficult to overcome the desire to use the latest, sexiest technology that gets the most attention in the media. Still, taking a very conservative approach to choosing battery chemistry can be prudent. Durability and reliability under all foreseeable circumstances of weather and operational demands, together with a realistic determination of the life span of the batteries, will give a more detailed forecast of the power plant's financial performance than will speculation based on some new type of battery.

Another option is the use of high-energy capacitors. However, there are significant problems associated with bringing capacitors onto an energized AC bus if the timing isn't perfect for each phase (ringing). Using capacitors or even supercapacitors on the DC side of things can be a useful—though short duration—energy storage option. Essentially these capacitors act the same way a battery does but have a greater capacity for current outflow.

Making a model or using the SAM model from NREL will help determine which technology to use. Be sure to include as parameters the depth of discharge, charging and discharging rates, whether it will be a deep-cycle battery or one kept on a float charge, and how all of those factors affect the net battery capacity and life span. Also include the resulting footprint and housing requirements (including ventilation) and the battery support structures (racks), which may be required to have seismic endurance characteristics.

Mechanical energy storage

Flywheels represent another energy storage option, although the energy is stored mechanically rather than chemically or electrically. Anyone who has picked up their bicycle and given the front tire a spin and then given that tire just a periodic little burst of energy to keep it rotating will immediately grasp the idea of a flywheel. These are spinning discs or cylinders of heavy material, rotating in a vacuum and suspended by magnetic levitation to minimize all friction and windage losses. Consider a 5,000-pound disc of steel rotating at several thousand rpm and the time it would take to slow down (while generating electricity the whole time): Some models claim to last four hours. Other units made of lighter materials spin at much higher speeds (even 100,000 rpm) but do not provide power for as long as the metal ones do. These spinning bodies will require some energy input to get them going, but they also require small doses of energy to keep them spinning at the desired speed and another small parasitic load to establish and maintain the vacuum.

Connected to an electrical system, the stator is given periodic pulses of energy to keep the rotor spinning at the desired nominal speed. When power is needed, the system is used to generate electricity during the time it takes the rotor to slow down. This type of storage has the advantage of having an almost infinite number of charge/discharge cycles, limited only by the mechanical systems, not the electrochemical ones. There are a variety of sizes and masses to choose from, yielding a backup source of power for everything from a few seconds to a few hours. Also, although each flywheel may be small, often under 10 kW, they can be compounded together to generate much larger capacities. If short-term energy boosts and/or some kind of frequency stabilization are required, flywheels can be an attractive option if financed over the life cycle of the power plant.

Another form of energy storage that has virtually limitless charge/discharge cycles is compressed air. Compressed-air energy storage systems use a motor-driven compressor to pressurize the air inside a tank. When energy is needed, the process is reversed, and the pressurized air is allowed to bleed off through the compressor, using it as a turbine, to operate the compressor's motor as a generator. In theory, the capacity of this kind of energy storage is limited only by the number of tanks used and the pressure the air is brought to; however, in reality, it's not quite that simple, especially when applied on a large scale.

One problem with compressed-air storage is that the air will get hot as it is compressed, just as it does when using a turbocharger on a reciprocating engine. At least one company building these kinds of storage devices inserts a stream of fine water mist into the process. The water then absorbs the heat from the compressed air before being filtered out of the airstream. In this way, the power plant now has compressed cool air and a supply of hot water. Moreover, when energy is needed, hot water (from any source) can be used to mitigate the freezing effect of the release of the high-pressure air, effectively reversing the entire process.

Clearly, air storage poses none of the typical hazmat restrictions or spill protocols as would be needed with a flow battery, but certain hazards associated with pressurized systems must be addressed as part of the overall safety protocols for the power plant. Again, if the footprint will allow the space for multiple air tanks, using a compressed-air system can be a viable option for on-site energy storage, especially since there is virtually no limit to the number of charge/discharge cycles.

Other energy storage technologies exist. These include pumped storage hydro, which is on such a large scale and applicable only to specific geographical locations that it isn't practical for the typical on-site power plant situation. Similarly, another option, the advanced rail energy storage (ARES) system, isn't practical for most distributed generation applications. ARES uses a locomotive to haul a trainload of rocks along a track that heads up the side of a mountain. When electricity is needed, the train is allowed to descend by gravity, where it generates electricity in the same way a hybrid car does when braking that is passed to overhead electrical cables.

Even though there are alternative ways to store mechanical energy, the most common in use in the power industry remain flywheels and compressed air. Unfortunately, methods such as pumped-storage hydroelectric and ARES are too geography specific; in other words, both have their place, but the alignment of the required geography with the location of a facility in need of an on-site power plant would be pretty rare.

Thermal energy storage

When most people hear about thermal energy storage, they automatically think of the hot side of things—for example, large insulated tanks of hot water or molten salt. The reality is that there is just as much a need for energy storage on the cold side. Sometimes, the customer doesn't need high temperatures, but merely a reservoir of water at 80°F. In any case, there are several ways to go about it, and these ideas should be considered when planning a new on-site cogeneration power plant.

In most cases, there won't be a need for very-high-temperature thermal storage, such as the molten salt (solar salt) used in CST power plants. Insulated tanks, either above- or belowground, are all that is usually necessary to keep a given working fluid at some elevated temperature. If the upper limit on that temperature is in the ambient range, then the aboveground tank(s) can be painted black to make use of some passive solar thermal heating, instead of using heavy layers of insulation. Other methods include putting large boulders inside tanks. Once heated, these rocks will retain the heat. If you've ever stepped on a rock from the previous night's campfire, then you know firsthand just how well some rocks are able to retain their temperature.

On the frigid side, ice is a very straightforward source of cooling. For example, at a facility in a large city in the American Desert Southwest, a chamber the size of a two-car garage is filled with water, and then late at night, when the price of electricity is low, that water is turned into ice. Many piping elements run through the ice block so that, during the daytime, water passing through the pipes is chilled by the surrounding ice. This cold water is then plumbed to several buildings in the downtown area to be used as part of their air-conditioning systems. Even if it's not iced, an insulated tank of cold water at 40°F or less can serve as a reservoir of chilled water to supplement the on-site chillers during peak temperatures during the summer. Replenishing the chilled water in the reservoir can happen through the night, when demand on the facility's chillers is normally low, thus making it ready for the following afternoon's heat.

N? Or *N* + 1?

Once you've completed the analysis to determine how much generation and energy storage capacity you need, another factor to consider is installing just enough (N), or installing some spare capacity ($N + 1$ or more) to account for maintenance. In cases where reliability of power is the driving factor, perhaps you might want to consider $N + 2$. An analysis must be done to weigh the benefits of the increased continuity and reliability of power, at the excess costs, compared to any problem of insufficient power at critical times for the customer.

There are two ways to look at this, and the final project usually incorporates some combination of the two. One way is to separate the electrical loads of the customer into vital and nonvital load centers, such that in the event of a power supply shortage (e.g., a genset trips off-line), the nonvital loads can be shed, thus lowering the net electrical load to within the reduced capacity of the power plant. The other way is to look at the situation from a financial perspective and compare the cost impact of an outage to the cost of mitigation. This will depend on the number of times each year such a power loss can be expected—and the magnitude of the financial impact of each occurrence.

Similarly, consider whether to go with a single unit or multiple smaller units for a given total generation capacity. There is an economy of scale such that larger units cost less (on a $/kW basis) than smaller units. Again consider whether components will be single-point safe. Clearly, adding multiple engine generator sets spreads the risk of the failure of any given one of them. A balance of costs and benefits can be achieved by using multiple, smaller cogeneration units and spreading out the risk. This is especially true in markets such as California, where demand charges can be more than 50% of the customer's electric bill; the cost impact up front of having multiple smaller units and the economy of scale are offset by the money saved by mitigating the increased demand charges due to loss of generation capacity.

6

Distributed Generation Performance and Cost Analysis

Spreadsheet Models

Over the years, I have developed many spreadsheet models to try to figure out what my power plant designs would do over the course of their first year of operation (and beyond). With each new model, I added another new feature until I had included everything I could think of (and there's always room to add more), until my spreadsheets extended beyond column "DG" and were 35,065 lines long.

In this chapter, I will describe how I make my models. Because some of my calculations are proprietary, in these instances, I will tell you what I am doing and why, but you'll have to come up with your own formulas (sorry). Using data from existing power plants, I have validated these models by comparing their predictions to real outcomes.

Many factors govern how any given power plant will operate, and if these factors are not addressed individually, it can be difficult to analytically determine precisely why an existing plant is not operating as well as expected. For new power plants, or those that are designed but not yet built, failing to account for these factors will lead to guaranteed disappointment in the performance of the power plant (if it is ever built). Creating a detailed performance model will yield actionable results that will not only save you time and money but also provide a better product to your customer.

I use individual models for each different power and heat source, but they all include a provisions for a solar component, energy storage, and a prediction of the weather. Breaking down my model into specific areas of interest, I include sections for each 15-minute time interval covering:

- Time, date, and day
- Random numbers for variables
- A prediction of the weather
- An estimate of the net peak output of the generator(s), based on the weather

- The electricity imported by the prospective customer
- When to start an engine and load its generator
- Tracking the engine's load in specific ranges for maintenance purposes (low, partial, or full load)
- Using a protocol to unload the generator and shut down the engine if low load conditions are sustained until the desired threshold is reached
- Tracking the number of starts and stops of each engine
- Following the energy flows of the energy storage system, including the state of charge of the battery, number of charge/discharge cycles, among other parameters
- Calculating the new amount of power imported from the utility with the on-site power plant on line
- Determining the amount of fuel consumed by the engines
- Calculating the costs of the electricity for both the electric utility and the on-site power plant, as well as net savings to the potential customer

I then compare the results to the given on-site power generation problem. Ambient weather conditions end up playing a huge role in the performance of any combustion-based power plant, and the failure to account for weather-related effects at the beginning is the primary cause of disappointment at the end. Another variation to consider is using gensets in a peaker plant configuration as opposed to parallel operations. The benefits of each method will depend on the nature of the loads of the customer (and changes in those loads over the course of any given day). In any case, if you configure your model to allow for different operating modes, you will have additional information when comparing the various power plant configurations that meet your energy needs.

Given the size and complexity of the model described here, it will be helpful to create a spreadsheet and populate the cells with at least the column titles used with a note describing the purpose of those data. This can then serve as a template for future use.

Generic Model

The model I describe below is for a microgrid with three reciprocating engines and an optional solar (PV) field and an energy storage component (a battery). The first bits of the model are details I include in all my models, and the microgrid-specific bits will come into play later. I start the data-crunching parts, the meat and potatoes of the model, at line 15 or so. This allows room at the top of the sheet for defined data entry cells and summary information copied from the bottom of the sheet in a single place (e.g., annual engine run hours). Along line 15, I label the columns, and I leave line 16 blank so that I can insert preexisting conditions (e.g., whether the engines are running and the initial battery charge) before the first analytical line of data (which starts on line 17).

I usually start these models with three separate spreadsheets (arranged as tabs within a single file), between which data can be shared. These sheets are labeled "Model," "Weather," and "Comparison." Since the weather profile affects everything else, I start with that, after the reference information that defines each 15-minute interval given in columns A—C.

Column A is the time reference, based on 15-minute intervals, since that is the interval the data from the potential customer's electric utility is in (usually). I label cell A15 "Time" and A16 "0" (or leave it blank). Cell A17 is "15"; then A18 is "30," A19 is "45," and A20 is "1." This continues with A21 being "15" through A24 being "2" and so on, all the way down column A to the end of the day, with cell A108 being "23" and A112 being "24." A113 starts another day as "15" and the process continues down column A through the end of the calendar year with cell A35056 being "24."

Column B is the date reference, with B15 labeled as "Date." I leave B16 blank, as the "0" time reference in A16 is actually the end of December 31st (i.e., "31 DEC," or whatever day/month designation you prefer) of the previous year. B17, at the 00:15 time stamp in column A, is the first actual data point in the New Year, which is "01 JAN." B113 becomes "02 JAN", aligning with the "15" in cell A113, and this is continued through to the end of the year.

Column C is the day reference, with C15 being "Day." Having the day of the week as a reference is very handy, as it is helpful to differentiate weekdays from weekend days (as well as denote holidays). Most electric utilities have different rate structures for the different days of the week, and these will be incorporated into the economic section of the model, which is described later in this chapter (see the "Financial Model" section). Using a calendar for whatever year of data you're analyzing, insert the appropriate day in cell C17, C113, and so forth, all the way down to the bottom of the sheet. These first columns are shown in figure 6–1. Also note that, although not shown in this example, I highlight the weekend and holiday data in bright color to make it easy to notice when scrolling through the data set.

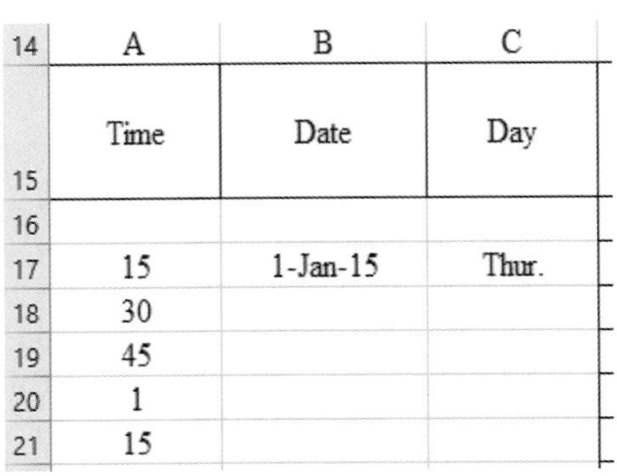

Fig. 6–1. "Time," "Date," and "Day" columns

Weather

Where you put the weather information in your model is not important—what is important is simply that you have a weather section, because the ambient conditions on any given day will have a significant impact on the performance of the power plant. If you are planning to enclose your power plant in a temperature-controlled environment, this bit of the model becomes very simple. However, if your power plant will be exposed to the ambient outdoor temperature, even if protected from the elements, the impact of the weather on the operation of the equipment must be taken into account. Most airports keep a log of the local weather data; conveniently, it's common to find these data in 15-minute intervals. Digging through the data on the weather.gov website can reveal a treasure trove of useful information.

As a bare minimum, include the daily high and low ambient temperatures in your model, because these factors have tremendous impacts on the performance of any power plant that uses the combustion of a fuel as the primary source of power. I also include cloud cover predictions to model the amount of solar radiation reaching the ground; these calculations may sound like voodoo, but bear with me as I describe the method behind my weather predictions. Thanks to the Internet, I can usually find the following information for the site I am evaluating (or local data close enough to the site), often as a graph of the annual data:

- Median cloud cover (percentage)
- Cloud cover types (clear, mostly clear, partly cloudy, mostly cloudy, overcast)
- Probability of precipitation (percentage)
- Types of precipitation (drizzle, light rain, moderate rain, heavy rain, thunderstorms, snow)
- Relative humidity (percentage)
- Wind speed
- Fraction of time spent with various wind directions
- Hourly temperature bands

Cloud cover

From the website weatherspark.com, I have adapted a graph of cloud cover types versus months of the year, given as a probability (percentage) for the location being analyzed. Within it, I divide each month into quarters and obtain the percentages for each type of cloud cover for each quarter-month period. Making any given month equal to four weeks allows easier interpretation of the graphical weather data. I put this information into a new spreadsheet ("Weather"), as a tab on the model, which will ultimately be only one of many separate spreadsheets that make up the performance model. This information is shown in table 6–1.

Table 6–1. Cloud cover probabilities

Month	Quarter	Overcast	Mostly Cloudy	Partly Cloudy	Mostly Clear	Clear
1	**1**	46	57	68	75	25
	2	45	56	67	75	25
	3	45	56	67	75	25
	4	43	55	66	75	25
2	**1**	40	53	65	74	26
	2	38	51	63	72	28
	3	34	48	61	71	29
	4	31	45	59	69	31
3	**1**	28	43	58	68	32
	2	27	42	56	67	33
	3	25	39	54	65	35
	4	24	37	53	64	36
4	**1**	21	36	49	61	39
	2	19	34	48	59	41
	3	17	30	45	57	43
	4	15	28	44	56	44
5	**1**	14	25	40	53	47
	2	13	24	38	52	48
	3	10	20	34	47	53
	4	8	17	30	44	56
6	**1**	7	15	27	40	60
	2	6	14	25	38	62
	3	5	12	23	36	64
	4	4	10	21	34	66
7	**1**	4	9	19	32	68
	2	3	8	19	30	70
	3	3	8	19	30	70
	4	3	7	18	30	70
8	**1**	3	7	18	29	71
	2	3	8	18	29	71
	3	4	8	19	29	71
	4	4	9	19	30	70
9	**1**	4	10	20	31	69
	2	4	10	21	31	69
	3	5	12	23	33	67
	4	6	13	25	34	66
10	**1**	7	16	27	37	63
	2	10	19	30	40	60
	3	14	24	36	46	54
	4	17	28	39	48	52

Month	Quarter	Overcast	Mostly Cloudy	Partly Cloudy	Mostly Clear	Clear
11	**1**	20	30	43	53	47
	2	23	35	47	56	44
	3	30	41	54	62	38
	4	32	44	56	64	36
12	**1**	38	49	60	68	32
	2	40	50	62	70	30
	3	44	54	66	73	27
	4	45	56	67	74	26

Estimating as best I can, I also generate a probability distribution of the amount of solar direct normal irradiance (DNI) that can penetrate the given cloud layer (as a percentage of the total possible amount) using a defined high and low limit. This is shown in table 6–2.

Table 6–2. Range of DNI penetration

	Overcast	Mostly Cloudy	Partly Cloudy	Mostly Clear	Clear
High	30	50	75	85	100
Low	15	20	40	60	85

Even though I have taken a few meteorology classes, I am certainly not a meteorologist; nevertheless, the values in table 6–2 are derived from personal observations, estimating the amount of sunshine that actually reaches the ground (or a solar [PV] panel) over the course of a year. Even on a "clear" day, there can be enough haze in the air to act as a filter, attenuating the DNI by up to 15%—or even more if you're analyzing a site in (or near) a major metropolitan area, such as Beijing (for an extreme example). This is why the "Clear" column has a range of 85%–100%. Thus, although the values in table 6–2 represent my estimates of hourly averages of the DNI conditions on the ground for each sky coverage for central California, you should tailor your values to match your given location.

In any case, the values in table 6–2 reflect my estimates of the range of available sunlight reaching a PV panel or a parabolic trough given the rough category of cloud cover predicted from the random number generator. Yes, this method of prediction may seem like voodoo, but at the end of the year, the averages will have worked themselves out with offsetting errors, yielding a net result that should be fairly close to the real conditions (even those yet to be experienced).

Randomness

To keep things simple, I treat the weather at any given time as a random variable, but within limited ranges to cover an hourly average; there's not much point in trying to be more precise than that (similar to dealing with significant figures in engineering calculations). I don't think it's reasonable for the model to allow for a fully overcast sky in one 15-minute interval and a perfectly clear sky in the very next interval; however, an hour after initially overcast conditions, the weather could be completely different. If you've ever lived in Denver, you may have heard the expression, "If you don't like the weather in your front yard, go to your back yard—it'll be different by the time you get there." The basic idea is to use the spreadsheet's built-in random number generator to give you a value as a starting point, assigning a cloud cover amount to that value and then using the random number generator again to yield a figure that is between the two limits of net DNI available for that particular cloud cover assignment.

Looking at the first quarter of the first month of the year in table 6–1, we see that there is a 46% chance that it is overcast. The values in the table are cumulative. Thus, given the 46% chance of overcast conditions and an 11% chance that it's mostly cloudy, there is then a 57% chance that it's either mostly cloudy or overcast. Furthermore, you could include the 11% chance that it's partly cloudy and the 7% chance that it's mostly clear, to round things out; subtract that total from 100%, and what's left is the chance that there is a clear sky that day (25%). From this, we can allow a randomly generated number between 1 and 100 give us what the sky coverage is (based on what range the number falls into), and then a second randomly generated number can be used to determine the range of attenuation for each category of sky coverage to give the actual (predicted) percentage of the incident DNI reaching the ground or the PV panel.

Along line 15 in the model, starting with column A, we now have "Time," "Date," and "Day," so cell D15 can be "Random." From cell D17 (D16 is blank) down to the bottom of the data, the cells are "=RANDBETWEEN(1,100)." These values will change every time the return key is hit, which has its purpose, but it's also good to have a fixed set of randomly generated numbers for analysis, so I label cell E15 "Fixed." To populate this column, simply copy the values in the "Random" column (D) and copy into the "Fixed" column (E) as either values only or values and number formats. Cell F15 is "Wx" (weather), and for the values in this column we must reference the weather table (table 6–1) on the "Weather" tab, which occupies columns A—G on that sheet.

For example, for the first quarter of January (approximately seven days), I use the following formula for determining the sky cover in each cell in column F:

=IF(E17>Weather!$F$2,Model!$K$15,IF(Model!E17>Weather!$E$2,
Model!$J$15,IF(Model!E17>Weather!$D$2,Model!$I$15,IF(Model!E17>
Weather!$C$2,Model!$H$15,$G$15))))

In other words, I compare the randomly generated number in cell E17 to the cumulative percentages in the weather table, as shown in table 6–3.

Table 6–3. Cloud cover, the first week of January

Month	Quarter	Overcast	Mostly Cloudy	Partly Cloudy	Mostly Clear	Clear
1	**1**	46	57	68	75	25

If the randomly generated value is above 75, it is a clear day. If not, then is it above 68? If so, then it must be a number from 69 to 75, and it's mostly clear. If it is not above 68, is it above 57? If it is, then it's a partly cloudy day, and if it's not, then is it above 46? If it is, then it's from 47 to 57 and therefore mostly cloudy. Finally, if it is 46 or below, then it is an overcast day. By this method, every 15-minute interval will have its own randomly generated cloud cover, based on a set of limits for that particular week (or quarter-month), throughout the year.

To make things a bit more realistic and to smooth out some of the extremes of the randomness (e.g., overcast for 15 minutes, clear skies during the next interval, then back to overcast), I average the results over each hour. For more column headings, I label the cells G15—K15 as follows: "Overcast," "Mostly Cloudy," "Partly Cloudy," "Mostly Clear," and "Clear." To estimate the solar DNI attenuation for each sky cover condition in cells G17—M17 (and continuing below, to the bottom of the sheet), I refer to the result in the "Weather" cell (F, Wx). The first several 15-minute intervals are in table 6–4.

Table 6–4. Cloud covers

Random	Fixed	Wx
27	51	Mostly cloudy
76	12	Overcast
69	100	Clear
87	96	Clear
22	11	Overcast
5	40	Overcast
10	32	Overcast
45	42	Overcast

In the space above line 15 in columns F—K, I create a small table with the ranges of DNI attenuation factors that are associated with each of the cloud cover possibilities. Just because it's a clear day doesn't mean there will be 100% optical transmission through the atmosphere. Haze, smoke from nearby wildfires (which happens a lot in California), and other factors can still reduce the DNI reaching your PV panels, even though it's technically a "clear" day. (This was presented in table 6–2.) In other words, if the random number generator's input to the cloud cover table shows that a particular 15-minute period will have partly cloudy conditions, then the amount of DNI reaching the ground will be between 40% and 75% of the full amount possible.

The next step is to continue to populate the model with specific DNI attenuation factors that will be between the high and low limits established in table 6–2. There should be only one entry, with the other cells being a zero, as shown in table 6–5.

Table 6–5. DNI attenuation examples

Wx	Overcast	Mostly Cloudy	Partly Cloudy	Mostly Clear	Clear
Mostly cloudy	0	23	0	0	0
Overcast	18	0	0	0	0
Clear	0	0	0	0	93
Clear	0	0	0	0	97
Overcast	19	0	0	0	0
Overcast	22	0	0	0	0
Overcast	27	0	0	0	0
Overcast	19	0	0	0	0

In column L, I summarize the weather factor (denoted "Wx Factor"). Cell L15 is labeled with this column heading, and each of the cells below, to the bottom of the data in column L, is simply the sum of the cells in columns G—K. Since only one of those cells should be populated, this also makes for an easy cross-check that the model is behaving as desired. For example, cell L17 is:

=SUM(G17:K17)

So now we have an estimate of the random nature of the weather for each 15-minute interval for the entire year. Because the weather usually doesn't change from fully overcast to absolutely clear skies as quickly as 15 minutes, I attempt to smooth the differences among the data into an hourly average. Over the course of a full year, this helps to smooth out the outlying data points into a more realistic representation of natural weather variation—and its impact on solar power sources. This is shown in column M (labeled "Wx Fx Avg."), and it is simply the average value of each block of 4 interval data, with cell M20 being the first populated cell (table 6–6).

Table 6–6. Weather and DNI attenuation

D	E	F	G	H	I	J	K	L	M
Random	**Fixed**	**Wx**	**Overcast**	**Mostly Cloudy**	**Partly Cloudy**	**Mostly Clear**	**Clear**	**Wx Factor**	**Wx Fx Avg.**
15	51	Mostly cloudy	0	47	0	0	0	47	
37	12	Overcast	20	0	0	0	0	20	
72	100	Clear	0	0	0	0	100	100	
19	96	Clear	0	0	0	0	99	99	66.5
84	11	Overcast	28	0	0	0	0	28	
29	40	Overcast	21	0	0	0	0	21	
85	32	Overcast	21	0	0	0	0	21	
93	42	Overcast	24	0	0	0	0	24	23.5

I learned a long time ago that, even though the formulas are pretty simple and the odds of an error are low, it is practical to insert cross-checks into every spreadsheet model. There are many ways to do this, and the built-in formulas in most spreadsheets are helpful (especially when combined). In this case, since only one cell in columns G—K should have a number greater than zero, one way to verify results against expectations is to use a column labeled "Check" (e.g., column N), and have the cells in this column use the following formula (e.g., cell N17):

=COUNTIF(G17:K17,">0")

Since only one cell in the G—K range should be above 0, the result should be a 1 in each cell in the N column. At the bottom of the data set, another check can be made to verify that there are no cells in the column that have a number above 1 and another to verify no cells are below 1. When the results show anything unexpected, the use of conditional formatting of the cells in the area of concern (to highlight the errors) makes it easier to find them. Now, the forecast of actual DNI reaching the ground is complete. When some solar energy is considered in the hybrid mix of power and heat sources, a reasonable estimate of production can be made—and it will be a lot less than what the nameplate information predicts.

Temperature

The ambient temperature at any given moment has a tremendous impact on the performance of the equipment used in a cogeneration power plant. Consider the effect of air temperature on the heat rate of an engine or combustion turbine, as well as the follow-on effect of changing the costs associated with the fuel purchases; the effect of the difference (delta) between the temperature of the air and of the fluids flowing through the dump radiators and the amount of heat transfer taking place; and the derating of power sources and the ensuing effects of lower energy generation (and therefore lower revenue). Furthermore, operations at a lower power level can reduce exhaust flow and temperature, with consequences on steam pressure in HRSG (and therefore steam flow), affecting both the customer and the generation of revenue from steam sales.

Fortunately, the local weather data for whatever site you're analyzing is readily available, even if it's just the high and low temperatures of each day of the year. If higher-resolution data are available (some airports log data every 5 minutes), then I would use those data in my model. If all I have to work with are daily highs and lows, it just means a simple formula needs to be used to estimate the changes in temperature over the course of each day. As with cloud cover, temperatures can also change quite rapidly; however, to keep things simple, I use a linear model for the change in ambient temperature over the course of each day. Over the course of a year, the outlying data points are averaged, but this is not unreasonable—and if all you have are daily data, that'll do just fine. Armed with just the high and low points, I look at the data qualitatively and make an estimate of the average time of day for each extreme. This will vary, as the true high temperature achieved during any given day will occur at a different time depending on the local conditions.

I start with labeling the next three columns, in cells O15, P15, and Q15 as "High Temp.," "Low Temp.," and "Air Temp.," respectively, and insert the daily high and low temperature data in the same row as the first electrical load data (the 00:15 time interval) for each day of the year (see table 6–7). This will be a different time of day, for both data points, peculiar to the location of the site. For central California, on an annual average basis, I chose 4:00 p.m. for the high-temperature point and 5:30 a.m. for the low-temperature point. Using these two times of each day as anchors and knowing the high and low temperatures for each day, I simply use a linear equation to go from one day's high to the next day's low, and the same for the low temperatures rising to the next high point. In this way, I have a temperature estimate for each 15-minute interval for the entire year. Making a graph of these data can help with a qualitative comparison to a graph of the actual high and low temperatures from historical weather data, to make sure there are no serious deviations from the normal temperature distribution over the course of a year. Even though the weather of the past does not constitute a 100% accurate forecasting tool to predict the weather in the future, you can use historical averages instead of any particular year's data. The goal is not to become an accurate weather forecaster for the sake of knowing the weather; rather it is to use an average weather pattern, in this case the temperature distribution, to predict the impact that those temperatures will have on the equipment used in the power plant. Any weather-related modification to the nameplate performance characteristics will provide a more realistic prediction than just using the nameplate information alone. Over the course of a full year, any significant variation pertaining to any specific day of the year will be absorbed.

Table 6–7. Temperature profile cells

N	O	P	Q
Check	**High Temp.**	**Low Temp.**	**Air Temp.**
1	50	31	31

With the daily temperature profile estimated, the next step is to determine the maximum power the electrical generator(s) can produce during each 15-minute interval. In this case, the model is for a cogeneration power plant using reciprocating engines. For any given engine-generator set, the OEM provides a simple document with the basic specifications of the unit, including the performance characteristics under nominal conditions. The more detailed versions of these specification sheets also include tables or graphs of the effects of ambient temperature, altitude, and fuel quality on these performance characteristics. By creating either a look-up table or equations for the given curves, you can determine the effects of the ambient air temperature on the engine—that is, the derating—and come up with the new, temperature-adjusted maximum output of the generator. For some engines, these curves are almost linear, which makes it easy; for others, you may have to interpret the graphs provided and come up with your own interpolation. An example is shown in table 6–8. The generator output is in kilowatts, and the heat rate is in units of BTUs per kilowatt-hour.

Table 6–8. Effects of temperature on engine-generator performance

Q	R	S
Air Temp.	**Max Generator O/P**	**Heat Rate**
31	375	10,960

With the cloud cover data still close by, the next two columns are for the useful sunlight reaching a PV panel and the net kilowatts of AC power that the array of PV panels and inverters can produce. There are many sources of information to determine the DNI for your specific location (usually given in units of watts per square meters, denoted "W/m^2" in the spreadsheet). The maximum DNI must be attenuated by the given weather conditions, as predicted by the model, to yield a net DNI that can reach the PV panels. Naturally, the choice of PV panels and the size of the array will be needed as inputs, as well as the specifications of the inverters used. Inverter efficiency is also something to consider, in that there will not be a 100% transfer of the PV-produced DC energy into AC electricity. Again, once the maximum output of the PV system is estimated, the attenuation due to the weather can be accounted for. This will yield a reasonably accurate prediction, over the course of a year, of the contribution of the given PV-based power system to the hybrid power plant. This is shown in table 6–9.

Table 6–9. Solar (PV) contribution

T	U
Direct W/m^2	**PV kW AC**
856	550

Now we have something in row 15 for columns A—U. The time of day, date, day of week, and weather data are being used, so next let's focus on the electrical side. Cell V15 I label "Import," and I populate column V with all of the available 15-minute interval data from the prospective customer's facility that aligns to the date and time references. Remember that these data are usually in units of kilowatts (rate) and not kilowatt-hours (quantity).

So far, my models for the various types of power and heat sources have been pretty much the same, but now a differentiation is introduced. Although, theoretically, you could create one master spreadsheet model that covered all of the available power and heat sources out there, it would be so huge that the calculations would probably bog down the central processing unit of the computer being used. Even as it is—with individual models for reciprocating engines, combustion turbines, microturbines, and so forth—if I have too many spreadsheet models open at one time, my computer lets me know it isn't happy by slowing to a crawl.

I will continue the model with the idea of the power plant using three reciprocating engine-generator sets and an energy storage component, all operating in parallel with the electrical utility. Reciprocating engines present a special challenge, especially in a modular design using multiple units. Specifically, I'm referring to the limits on low-load operations. How low do you let the generator's load go before reaching the threshold of operational pain and turning off the engine that's driving it? How do you share load among the generators to maximize efficiency? How does this plan affect the overall reliability of the facility? In other words, although running one engine-generator set at 100% load is more efficient than running two units each at 50% load, what happens if that one engine trips off-line owing to a fouled spark plug (or any other anomaly)? What about throwing some energy storage into the mix?

The idea that a battery-supplied inverter (that's been floating on the main generator bus) could instantly swing into action and carry the total plant load until one or more engine-generator sets can be brought on line is a fantasy. Even with an inverter rated to carry the full electrical load of the facility, inverters cannot go from zero to fully loaded instantly. There is a finite ramp ability, where the rate of change of the output of the inverter is going to limit its ability to provide power to the bus (at first). If the facility was operating in island mode at the time, the protective relays could sense the trouble and trip off the inverter long before it could provide enough power to the bus. Even if it were to stay on line, the voltage dip could be severe enough to cause other problems subsequently.

Using the space above the 15th row in columns V—Y, I make a selection grid where the size of the individual engine-generator units (and all the derivative information) can be used as reference points in the rest of the model. This is shown in table 6–10.

Table 6–10. Engine-generator operational information

No. of Units	1	2	3
Full kW	375	750	1,125
Starting	225	450	843.75
30%	112.5	225	337.5
50%	187.5	375	562.5

In this case, I'm using three engine-generator units, with each generator rated at 375 kW of continuous output (prime power, not backup power). Regarding the "Starting," "30%," and "50%" information, these are key operational points of concern. One question to ask is, What is the minimum electrical load before I allow the first generator to be started? In this case, I am using 60% of full load as a threshold; in other words, starting from 0, the total electrical load would need to reach 225 kW before the first engine can be started and the first generator brought online. Although you can set the starting threshold wherever you want to, I choose 60% in my model to achieve a low-load engine-shutoff threshold; that is, if an engine is operating somewhere between 30% and 50% of full load for three consecutive time intervals (or four or five intervals), then the generator is taken off-line and the engine is shut down. This is to prevent extended operation at low loads (and avoid all the bad things that happen if it were allowed to continue). To start the engine at 50% load, without knowing whether load was going to increase or decrease after the starting threshold was reached, raises the concern for excessive start/stop cycles of the unit; some engines are fine with that, others, not so much.

Note that this is will mitigate, rather than prevent all together, the situation of hunting for the engine control system. That is, if the electrical load is not stable or sufficient enough, the unit will just shut right back down due to the protections against continuous operation at low engine loads. Setting the starting threshold higher (e.g., at 70%) avoids the low-load issue but at the cost of lost revenue—that is, waiting to start the engine-generator set means waiting to get the revenue stream on line, perhaps unnecessarily. This starting threshold is carried over to the two- and three-unit conditions.

Table 6–11 also shows the 30% and 50% kW load values of interest. The 30% figure is also a *kill threshold,* and any time the electrical load falls below that amount the engine is shut down.

Table 6–11. Engine-generator low-load information

V	W	X	Y	Z	AA
Import	**Load Eng. 1?**	**Fully?**	**>.3?**	**<.5?**	**.3-.5?**
	0	0	0	1	0
53	0	0	0	1	0

Thus, with the import data from the facility being analyzed now in column V, the first question, to be answered in column W, is whether there's sufficient load to start the first engine-generator set. This is a simple "IF" formula, comparing the value in the import cell to the reference cell, giving a logic 1 if true and a logic 0 if false, like this, with W3 being the reference threshold:

```
=IF(V17>$W$3,1,0)
```

The next information I need is whether the generator can be fully loaded. In column X ("Fully?"), I use the same formula to give me a logic 1 if the available load is greater than the full load ability of the first engine-generator set. I then use the next several columns to get information about the operation of the first unit—all I know now is whether to load the unit and if it is able to be fully loaded. Again using an "IF" formula, I first verify that the load will be above the low-load threshold of pain (30% in my case) and use column Y for that. Next, I check for the other threshold of low-load operations and whether the load is below 50%, which I specify in column Z. In the next column, AA, I start tracking the amount of time the load is between 30% and 50%, using a logic 1 if the values in columns Y and Z are both 1s. This is shown in table 6–11.

Two other points of interest are if the engine-generator is loaded above 50% and if it is loaded below 100%, and I note these data in columns AB and AC, respectively. I can then identify when the load is in the three power bands of interest, for O&M information, as "Low Load" (30%–50%), "Partial Load" (>50% to <100%), and "Full Load." Columns AE, AF, and AG respectively show that same load information in units of kilowatts (rather than relative percent of full load). I follow these three pieces of information with another cross-check, in that there should be only one bit of data greater than zero in any of the three load ranges (column AH). This is shown in tables 6–12 and 6–13.

Table 6–12. Engine loading bands

AA	AB	AC	AD
.3-.5?	>.5?	<100%?	.5-.99?
0	0	1	0
0	0	1	0

Table 6–13. Engine load amount for each load band

AE	AF	AG	AH
Partial Load .3-.5	**Partial Load .5-.99**	**Full Load**	**E1 Load Check**
0	0	0	0
0	0	0	0

To have it in one place and to make it easier later, I create a summation of the three load ranges (if the cross-check is satisfied), which should be the net load on the first engine-generator set, and I label this "E1" (column AI). Now I start the counting process to determine if the engine has been operating in the low-load range for the requisite amount of time to trigger a low-load shutdown using columns AJ and AK. This is also why I start the data stream at row 17, not 16—because I can insert the previous condition of the engine-generator set in row 16 for the columns used in determining any action resulting from a triggered event (e.g., a low-load shutdown). I use column AJ for the cumulative 3× criterion and column AK for the 4× criterion to trigger a low-load shutdown of the engine. If the load goes above the 50% threshold for even one 15-minute interval, it resets the clock for the low-load trip. As discussed in chapter 5, for mitigation of the problems associated with continuous low-load operation, the engine needs to be run at high-load (>80%) for at least as long as it was run at low-load levels. In this case, although operating at 51% will reset the trip wire for a low-load shutdown, the average operational time over the course of a full year will serve to protect the engine from excessive amounts of time operating at low loads. The 3× threshold is only 45 minutes of continuous low-load operation, which is very conservative.

At the bottom of the data set, a summation of the number of time intervals the engine-generator set operated in each of the three load ranges will help show the relative amounts of cumulative time spent in each range—hence, the maintenance aspect of collecting that information. I use column AL for a logic 1 if the low-load trip condition is met and column AM as a logic toggle to represent the engine's condition, with online being logic 1 and off-line being logic 0. Columns AN and AO are logical condition checks. Column AN is a logic *true* if the engine was previously running at a load above 30% with no low-load trip condition present (counter being active), otherwise it gives a logic *false*. Column AO gives a logic true if the engine was previously off line and sufficient load now exists to start the unit, otherwise it's false.

If you're keep track or making your own spreadsheet model as we go through this, you should now have the column identities summarized in figure 6–2. The reasons for having all of this information (if they aren't clear yet) will become evident as we get to the end of the model and distill the results into a summary of operational and performance metrics.

Fig. 6–2. Column identities for spreadsheet model

A	B	C	D	E	F
Time	Date	Day	Random	Fixed	Wx
G	**H**	**I**	**J**	**K**	**L**
Overcast	Mostly Cloudy	Partly Cloudy	Mostly Clear	Clear	Wx Factor
M	**N**	**O**	**P**	**Q**	**R**
Wx Fx Avg.	Check	High Temp.	Min Temp.	Air Temp.	Max Generator O/P
S	**T**	**U**	**V**	**W**	**X**
Heat Rate	Direct W/m^2	PV kW AC	Import	Load Eng. 1?	Fully?
Y	**Z**	**AA**	**AB**	**AC**	**AD**
>.3?	<.5?	.3-.5?	>.5?	<100%?	.5-.99?
AE	**AF**	**AG**	**AH**	**AI**	**AJ**
Partial Load .3-.5	Partial Load .5-.99	Full Load	E1 Load Check	E1	.3-.5 3x
AK	**AL**	**AM**	**AN**	**AO**	
.3-.5 4x	Low-Load Trip	Eng. 1	Cond. 1	Cond. 2	

Continuing onward, the next four columns are information about the status of the engine, and while not critical bits of information, they can come in handy at the end of the year. In column AP, I verify whether the engine is off-line, with a logic true if conditions 1 and 2 are both false. Also of interest is whether the engine is continuing to run, so in column AQ, there will be a logic 1 if condition 1 is true and a logic 0 if not. Similarly, in column AR, there will be a logic 1 if condition 2 is true and a logic 0 if not (restart). This makes it particularly easy to total up the number of times the engine was started during the year, which can affect the maintenance necessary to perform on the engine. Similarly, in column AS, I will get a logic 1 if the unit is off-line (if the cell in column AP is true), so at the end of the year, I can tally how much downtime there was—clearly, to maximize revenue, this number needs to be minimized.

After all that information is gathered about the first engine-generator set, we can conclude by finding the amount of electrical load (in kW) that the generator is supplying to the main bus. In column AT, if the engine is online, then the value of the load will be whatever the imported kilowatt amount is, up to the amount determined as the unit's maximum output (per column R). During high ambient temperatures throughout the year, this amount will be less than the nameplate data for the rated maximum output of the generator. This set is shown in figure 6–3.

Fig. 6–3. First engine-generator set data

AP	AQ	AR	AS	AT
Engine Off?	**Cont. Run?**	**Restart?**	**Off-line?**	**E1 Load**

If the amount of electrical power being imported from the utility is greater than the full load capacity of the first generator, consideration can be given to starting the second unit. This is where options come into play, and if you set up your model well, you will be able to test all the various combinations of equipment to determine the best answer for the given situation—and by "the best" I mean whatever it is you are optimizing for. Once the model is complete, you should be able to maximize or minimize any target value and plot the results; this, along with sensitivity analysis, will help you determine the best fit for your facility. For example, you can optimize for having a minimum amount of power imported from the utility, engine downtime, the number of engine starts, or some other objective. These results will vary with the operational paradigm you employ, and a good model will allow the analyst to choose from a menu of options and compare the various results. Are the engine-generator sets going to be operated as a peaker plant, where the engines start one at a time until all are online, as necessary—or will the plant be run with the units operating in parallel, sharing the available load when possible? What about the energy storage side of things: How and when will the stored energy be put to use? One way to figure out what to do is to make a model that optimizes each 15-minute interval for the metric of choice.

If you let your imagination run wild, you will see that to have one, all-inclusive model that covers all of the operating scenarios for using three reciprocating engines and some kind of energy storage capacity would end up being a monster. And while you'd be right, it is a monster that can be tamed. The key is to insert in your model far more cross-checks than you think you need. When the results of the data crunching don't make sense, these cross-checks will make it possible to find the error—and it can be easy to unintentionally insert an error when copying and pasting formulas than one might anticipate. Color-coding also helps make clear the associations among the data.

Microgrid Model

Now that we have the weather modeled and can predict the maximum output of the generators (compared to nameplate data), we can analyze a power plant configuration. Taking the model I have presented so far as is, I will add one method of figuring out how to employ the various power and heat sources used in a given facility—namely the microgrid design (fig. 6–4).

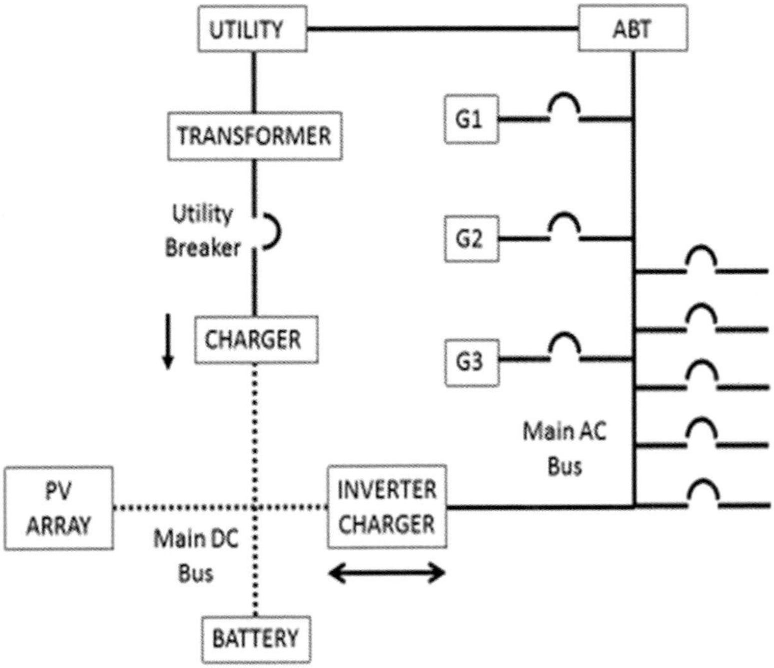

Fig. 6–4. Microgrid Design Concept

In a microgrid configuration, the utility supplies power through a transformer and intertie circuit breaker, providing AC power to a battery charger. The battery charger then supplies DC power to the main DC bus, which can charge the battery and/or feed the inverter-charger to supply AC power to the main AC bus. This way, no interconnection agreement with the utility is necessary, since there is no way for any of the on-site generation assets to supply power to the grid by using interlocks in the *automatic bus transfer* (ABT). Under normal conditions, the main AC bus is not connected to the utility. Under emergency conditions (e.g., loss of all AC power on the main bus), the ABT senses the need for emergency power and automatically connects the utility to the customer's electrical distribution system. Interlocks in the control system then disable the on-site power plant's sources of AC power, preventing any interconnection with the utility.

The battery charger and PV panel array provide DC power to the main DC bus, which is connected to the inverter-charger unit. The inverter-charger can work in either direction—using DC voltage to generate AC power to the main AC bus or using AC power to generate DC voltage—but cannot work both ways at the same time. The inverter-charger needs to be sized such that the maximum AC output of the inverter is at least as much as the starting threshold for the first generator. The size of the PV array is a variable to be determined later, but the size of the battery needs to be addressed up front. Other parameters of the battery system also need to be identified early, such as Ah capacity, minimum allowable charge level, and acceptable number of charge/discharge cycles (all to determine the life of the battery). There are three generators to supply AC power to the main AC bus, and these can be driven by any technology.

One method I use is to make a table of the range of available generation assets compared to what should be done under each set of conditions (in my opinion, anyway). In this case, when do we use the battery to supply the inverter, when do we use a generator, and when do we use a combination of both? What about when the battery is dead? When do we use the charger to recharge the battery, and when do we use the generators to make extra AC power to use the charger half of the inverter-charger? This can be handled by dividing the available resources—such as battery available and battery unavailable—into the logic of what resources to use and when. For this model, being a pseudo-off-grid microgrid, it will be optimized for minimal imported power from the utility. This decision matrix is illustrated in table 6–14.

Table 6–14. Power plant load matrix

<table>
<tr><td>kW</td><td>Max Import</td><td></td><td></td><td></td><td rowspan="2">Inv+Batt G1,2&3</td><td></td><td rowspan="2">Charge from Util.</td></tr>
<tr><td>1,125</td><td>G1,2&3 Load</td><td></td><td></td><td></td><td rowspan="2">Charge 3</td></tr>
<tr><td>843.75</td><td>G1,2&3 Start</td><td></td><td></td><td rowspan="2">Inv+Batt G1&2</td><td></td><td></td></tr>
<tr><td>750</td><td>G1&2 Load</td><td></td><td></td><td></td><td rowspan="2">Charge 2</td><td></td></tr>
<tr><td>450</td><td>G1&2 Start</td><td></td><td rowspan="2">Inv+Batt G1</td><td></td><td></td><td></td></tr>
<tr><td>375</td><td>G1</td><td></td><td></td><td></td><td rowspan="2">Charge 1</td><td></td></tr>
<tr><td>225</td><td>G1 Start</td><td rowspan="2">Inv+Batt</td><td></td><td></td><td></td><td></td></tr>
<tr><td></td><td>Min Import</td><td></td><td></td><td></td><td></td><td></td></tr>
</table>

From the bottom row of table 6–14 upward, the first range of possible operation is from some minimal imported power (even zero) up to the starting load threshold for the first generator set. We could supply the AC bus with power from the inverter-charger unit, itself being supplied DC power from the charger (powered from the utility), and we would have to do so if the battery were not available. However, with the battery available, we can use "Inv+Batt" to denote that we will use the battery and the inverter-charger to power the main AC bus. Keep in mind, when selecting which inverter-charger unit to use, that it will need to handle the starting surges from the loads the customer turns on as well as the surges when the customer turns loads off; otherwise, you'll need something to absorb the excess AC power on the bus as the inverter ramps down its output. Even if the customer's electrical demand never goes above this range, the utility can supply power to the main DC bus via the charger when (not if) the battery is depleted. The PV array and/or the utility can also supply DC power to recharge the battery. When AC power is needed but the demand is below the threshold for starting an engine-generator set, it will always be preferred to use the PV array and/or the battery to provide DC power to the inverter (rather than import power from the utility).

The next range of operation pertains when there is sufficient electrical demand by the customer to get above the starting threshold of the first generator but it is still below the maximum output of that generator. Again, there are two parts to this—when the battery is fully charged versus when it could use recharging—and that determination will vary wildly with battery chemistry, size, and other details. If the battery is fully charged or not yet discharged to the point where it should be recharged, then the first generator can supply the main AC bus while being buffered by the AC output of the inverter-charger. If the battery needs recharging, the generator's output can be increased to its maximum, and the extra power (beyond the needs of the customer) can be channeled through the inverter-charger to supply DC to the battery. This also allows the engine that's driving the generator to operate at a higher load than it otherwise would, thereby increasing its operational efficiency and producing a greater thermal output (which can be stored if it isn't used). This zone is denoted "Charge 1" in table 6–15.

The third range ("Inv+Batt G1" in table 6–15) pertains when generator 1 is fully loaded and the customer's demand is below the point of being able to start the second generator and share the load equally with generator 1—and have both units operating above the low-load operating threshold. Fortunately, we again have the inverter-charger unit to supply AC power to the main AC bus, just as we had at the start, and operations are very similar to that first range. One big difference is that now, for load transients, there are two generators (generator 1 and inverter) supplying the main AC bus, so the ability to handle surges is much improved.

Extend this logic upward through the load ranges and it should be easy to understand how the system works. Once in the uppermost range, when the customer's demand for power exceeds the capacity of all three generators, there are only three options:

- One is to use the available DC voltage sources (PV array and battery) to supply the inverter-charger and send AC power to the main AC bus as long as the DC sources can last.
- Once the DC sources have been depleted, the second option is to use the utility to supply DC to the main DC bus, so the utility can then indirectly feed AC power to the main AC bus. For this transition to work, the output of the charger must be increased as the PV and battery strength wanes to prevent any interruption or drop in the AC output of the inverter-charger.
- The third option is to simply reduce the customer's load to within the on-site generation capacity. This will be an economic decision more than an engineering one—the power plant will work either way—as the cost of the extra power imported from the utility is weighed against the needs of the customer to have the loads operating. Here, the temptation to use larger engine-generator sets must be checked. Even though having excess generation capacity installed would reduce the probability of ever having to import power from the utility, it also means spending more time operating those engines at less than full load.

Before we develop the model further for this power generation system, some key metrics need to be identified, which will be used as reference points throughout the rest of the model. All of these reference materials I put in the space above the main menu along row 15. The particular items of interest now are:

- Nominal battery voltage
- Depth-of-discharge limit
- Battery kilowatt-hour limit (a function of the depth-of-discharge limit)
- Minimum Ah left in the battery (i.e., after the battery is "dead")
- Total kilowatt-hour capacity of the battery
- Maximum AC output kilowatts of the inverter-charger
- Maximum DC output kilowatts of the inverter-charger
- Any desired constraints—for example, stipulating that the minimum inverter-charger AC kilowatt output must be greater than the starting kilowatt load threshold for the first engine-generator set. Another method to find the minimum inverter-charger AC size is to find the maximum imported kilowatt amount and then subtract the total generation capacity of the engine-generator sets.

Starting in column AU, I check to see if the available load (the imported power in kW) is less than the capacity of the AC output of the inverter-charger (using the reference data described above), with a logic 1 if yes and a logic 0 if no. I label cell AU15 as "Available Load < Inv O/P?" The next information I want to know is whether the total electrical demand is beyond the capacity of all three of the generators, so AV15 is "L > all 3?" with a logic 1 if yes and a logic 0 if no.

Continuing through the matrix of table 6–14, I then determine in column AW if the available electrical load is greater than the combined capacity of generators 1 and 2 but less than the starting threshold for the third generator (i.e., G1&2 < L < G123S). Next, in column AX, I check whether the available load is greater than the capacity of the first engine-generator set but less than the starting capacity for bringing the second generator online (i.e., G1 < L < G1&2S). In column AY, I verify whether or not the available electrical load is below the threshold of being able to start the first generator set (i.e., L < G1S). In column AZ, I include a useful bit of information that is not critical to the model—namely, the trend in the electrical demand and/or the loading of the AC power production units (increasing, decreasing, or steady).

In the next four columns (BA—BD), I identify which of the four ranges the electrical demand is in (with the battery available) that use the inverter-charger as an AC source: Inv+Batt; Inv+Batt G1; Inv+Batt G1 & G2; and Inv+Batt G1,2&3. A good cross-check to verify that all of your formulas are not in error (even if not necessarily good) is to verify that only one of these four columns has a logic 1 in it (but I don't do that in this particular model). The next four columns (BE—BH) are used to identify which of the battery charging options is available but not necessarily being used: Charge from G1; Charge from G1&2; Charge from G1,2&3; Charge from Utility. We still have four battery-unavailable conditions that need identifying, so columns BI—BL are used for this: Start G1?; Start G2?; Start G3?; and Use Utility? to recharge the battery (or not).

The next four columns have to do with the inverter-charger unit as an AC source. The first, column BM, asks the question "Is inverter AC desired?" A cross-check is used to make sure that a DC power output from the unit is not also desired (column CN). For cell BM17, this resembles:

```
=IF(AND(OR(BA17>0,BB17>0,BC17>0,BD17>0)=TRUE,CN17<1)=TRUE,1,0)
```

Even though inverter-supplied AC power may be desired, there are at least two sources of DC power to the inverter-charger unit. The next column, BN, checks to see if the battery is available for discharging. This checks to make sure the remaining battery capacity is above the low-level limit set in the battery parameter constraints that were defined earlier. It also verifies that the inverter-charger unit is not already being used to generate DC power (since it cannot produce both AC and DC power simultaneously). Column BO gives a logic 1 if the inverter-charger is in any of the combinations of desired AC output that previous logic hold. That is, for example, either inverter AC is desired AND battery discharge is available, or inverter AC is desired AND import from the utility is desired.

In column BP, I calculate the AC loading the inverter-charger unit is providing to the main AC bus. This checks all of the various conditions under which the inverter's AC power is desired. The formula can be something like the following:

```
=IF(AND(BA17>0,BO17>0)=TRUE,V17,IF(AND(BB17>0,BO17>0)=TRUE,V17
-$AY$6, IF(AND(BC17>0,BO17>0)=TRUE,V17$-AY$4,IF(AND(BD17>0,BO17>
0)=TRUE,V17-$AY$2,0))))
```

In any case, I follow this column with two cross-check columns (BQ and BR). In column BQ, I verify that these conditions do not exist simultaneously: the battery is unavailable, the plant is still making AC power with the inverter-charger unit, and the inverter-charger is not being powered by the utility. The second cross-check, in column BR, is to verify that the inverter-charger unit is not trying to make AC and DC power at the same time. If everything is okay, I now have a good estimate of the inverter-charger's AC power output. These may seem like redundant cross-checks, but I have learned the hard way that debugging a model can be a nightmare when I don't include a lot of cross-checks.

The next five columns (BS—BW) are devoted to identifying the electrical load being supplied by the first engine-generator set, and these cover the four areas of the power matrix involving G1 and the net load on G1. The next five columns (BX—CB) do the same thing for G1 and G2, with column CB being the total load for both G1 and G2. Columns CC—CG cover the same areas for all three engine-generator sets. I follow all this with a couple more cross-checks, just to verify that only one set of generator combinations is providing power at a time (columns CH and CI).

To keep from having to refer back to the beginning of the spreadsheet to see the data, I use column CJ to redisplay the original imported power data. In column CK, I calculate the "new" imported power from the utility. Ideally, this amount would be zero, but if the total demand were to exceed the total generating capacity of all three engine-generator sets long enough for the battery (and PV panel array) to no longer be available, then it will be more than zero. There will also be times when one (or more) of the engine-generator sets will be unavailable, because of either a fault or periodic preventative maintenance. Routine preventative measures, such as oil changes, can be included in the model by way of tracking the run hours on each genset (but this is not a part of this particular model). Similarly, you can insert all sorts of threshold-triggered changes to the available resources for the power plant to deliver the electricity and thermal energy to the customer.

Next we perform a unit conversion, to keep things simple. I use column CL to convert the AC kilowatt output from the inverter-charger into kilowatt-hours, aligning it with the conventional general measure of battery capacity, which I use column CM to determine. Since this is the start of a new year, I assume the battery is fully charged and ready to go and use an ampere-hour capacity from the reference data (250 kWh, or 50% of the initial 500 kWh capacity). This is because I have set the depth of discharge at the 50% level, so even though it is a 500 kWh battery, I only want to use the top half of it.

The remaining battery capacity can be more complicated to calculate than it first appears, since there are many factors competing for the battery's attention, but they won't always be able to get it. For example, there could be a logical request for the battery to provide DC power to the inverter-charger unit to help sustain the main AC bus, but the battery had previously been depleted and is now in a charging cycle. In this case, the charge would need to take priority, and the extra DC power would have to come from the utility via the charger unit—and, if the inverter-charger was making DC power, it would have to ramp down and then switch to providing AC power. Part of this is also the strong desire to keep a battery charge going once it has begun.

The next several columns are devoted to the DC side, similar in layout to the AC side. I start with checking whether DC from the inverter-charger unit is desired (column CN). Next, I check whether DC power from the utility is desired (column CO). In the third DC-based column (CP), I use a counter to identify all of the charging cycles the battery is put through. The number of charge/discharge cycles is a common life-cycle metric for batteries, and it varies greatly among the possible chemistries and physical arrangements available. It can also affect the maintenance requirements of the battery system, the total battery capacity, and, therefore, depth-of-discharge limits, among other parameters.

In column CQ, I verify whether the inverter-charger unit is making DC power. This also serves as a reference point for the prior 15-minute interval period; as conditions change, the inverter-charger unit's desired job can change, but it can't always shift away from what it was doing previously.

Columns CR and CS are used for calculating the DC load of the inverter-charger, in kilowatts and kilowatt-hours, respectively. Columns CT—CV are used for system status and cross-checks. CT checks whether a battery charge is in progress; column CU checks whether there is a simultaneous demand for both AC and DC from the inverter-charger unit; and column CV checks whether the kilowatt loading is above 0 for both sides at once. This last cross-check is because, even though there may be a demand for both AC and DC power from the unit, that doesn't necessarily mean such power is being supplied; however, the dual demand request does indicate a problem with the model.

Columns CW and CX calculate the AC electrical loading, first for the loading on the engine-generator sets and then the total AC being produced (to include the inverter). Knowing the total amount of generated AC power, I subtract that from the original demand data to get a remainder—which, if everything is working correctly, should be zero—in column CY. To count how many times this happens, I use column CZ to identify every time the remaining AC power is above zero. Now, for run hour calculations and other analysis, I use the next four columns (DA—DD) to identify when only G1 is running, when G1 and G2 are both running, and finally when all three generators are online. I follow that with another cross-check, because there should never be a condition where more than one of those criteria are met simultaneously.

Now we get to the first bits of financial data, in columns DE—DI. Knowing how many generators are online, and what the total generated AC kilowatt output is, I calculate the percent loading of each of them. In the area above this part of the spreadsheet (lines 1–14), I insert a table containing the fuel usage data from the manufacturer's information sheet on the specific equipment I'm modeling. From this data set of percent engine load, together with the brake horsepower (BHP) and the associated fuel use (often given in terms of BTU/BHP h), you can calculate a BTU/h rate of fuel consumption, which is easily converted into amounts of consumption during a 15-minute interval. Using the OEM data, I plot the fuel usage versus percent load, and the resulting curve is close to linear most of the time (but not always). I use this graph to get the equation for the curve, and then I use the individual engine load amounts for each 15-minute interval to calculate the total amount of fuel consumed in that period. From those data, in yet another column, I calculate the heat rate (in BTU/kWh) for each period; this is particularly useful when performing sensitivity analysis later, because it makes clear when the power plant is operating at maximum versus minimum efficiency. This value in column DI should be cross-checked with the heat rate value in column S.

The performance model described in this section helps the cogeneration power plant developer to design a facility where the net imported power from the utility is at a minimum—if the sizing of the generation equipment, battery, inverter, and other components is well balanced. Most industrial and commercial facilities will not be as completely isolated from grid-supplied AC power as this microgrid design is for a couple reasons. First, and usually the limiting factor, is the ratio of generator size to the size of the largest electrical load, relative to the starting surge associated with that load. Most motors, including those that drive compressors, will have an initial starting current surge that can be five to eight times the normal run current amount. Some of the newer, ultra-high-efficiency motors can have even higher starting current surges, even approaching 20 times the nominal run current. The big problem with a high starting surge when in isolated microgrid or island mode is where the current comes from. When a power plant is operated in parallel with the local electric company, the current can be supplied by the utility as the on-site generation equipment comes up to speed. Even with that utility-supplied power, there will still be a momentary dip in the bus voltage.

For example, if you're trying to start a 500-horsepower compressor that's being powered by a single 375 kW engine-driven generator, the voltage dip will be so severe that the protection relays won't handle it, causing the generator output breaker to open on undervoltage and/or underfrequency. To mitigate this effect, most off-grid, microgrid, and island mode—capable power plants have extra generating capacity built in. While this helps keep the power plant online during large electrical transients, it also usually results in the engines driving those generators operating at low efficiencies and drives up the initial cost of the power plant (i.e., $N + 1$ is more expensive than N).

A second reason why most on-site power plants are operated intertied with the local electrical utility is that they need continuity of power. As the diagram of the power plant configuration used for the performance model (fig. 6–1) shows, if there is ever an emergency, the ABT switch can supply AC power directly from the utility to the main AC bus. Regardless whether it's a dead-bus shift or whether the ABT control system needs to synchronize the AC bus frequency with the utility's frequency and match voltages (if there's an intertie agreement), it will take a finite amount of time to make the shift. During that time, the customer's PLC-driven equipment will go off-line and need to be reset, which can result in damaging consequences that make it an unacceptable risk. Although modern stationary engines and combustion turbines are highly reliable, they are not infallible, and it's just a matter of time before any given unit trips off-line or must be brought down for maintenance. Under these conditions, having the utility being a source of relatively reliable power is comforting to the chief executive officer of the typical industrial facility. Remember that if the ABT has the ability to synchronize with the utility, an interconnection agreement will be required. If, however, a dead-bus shift—only device is interlocked with the generator and inverter output breakers, then an interconnection agreement should not be necessary. That is assuming the utility approves of the interlock system, such that if the ABT were connected to the utility, then none of the AC generator output breakers could receive a command to close from the generator control system.

Financial Model

When there is some net imported power from the utility, or just when you want to have the information handy, a financial model can be made to show the effects of changes in the design of the potential power plant. In many parts of the country, the local electric company is not only charging industrial customers for the kilowatt-hours of power they consume but also levying a *demand charge* based on the highest rate at which the entity consumed their power. By analogy if you drive a car from point A to point B, the cost of doing so depends on both how fast you go and what your highest speed is—regardless of how long you were at that top speed. There will be a fee—in addition to the cost of the fuel, based on a certain $/mph amount (that is usually tiered, so the higher that top speed is, the more expensive it becomes)—that is concerned only with the peak velocity. The size of this fee changes over the course of a day, such that the rates charged are higher during the day than they are at night, relative to the amount of traffic on the highways. The more desired (i.e., on peak) the driving time is, the higher the fee will be.

Electric utility companies will often break up a day into zones for billing purposes, commonly defined as *peak, partial peak, off-peak,* and—the kicker—*anytime.* I have seen an electric company artificially keep its per-kilowatt-hour rate low by jacking up the amount they charge for peak demands, to the point that the demand charges were more than half of a customer's electric bill. This is where on-site power can come to the rescue, in that every kilowatt generated on-site is a kilowatt that doesn't have to be imported from the utility. The typical on-site cogeneration power plant can easily deliver a net energy savings to the host customer, both in saving dollars per kilowatt-hour over the electric utility and by generating thermal energy cheaper than the customer could do it themselves.

The cost savings to customers in markets like California can be significantly larger, in that in addition to reducing their electric bill by 10%–15% on the per-kilowatt-hour rate, the savings from the avoided demand charges can approach 100%, which can be over half their bill. For example, let's take an industrial facility in central California, using their electric bill from August 2015 (the hottest part of the summer). During the month, this facility used 1,089,989 kWh of electricity, which cost them $102,909 (or an average of 9.52 cents/kWh). Their highest demand for that electricity was 1,884 kW, and the total of all their demand charges was $64,011 (or $33.98/kW), which is 62.2% of their charges for consumption and 38.3% of the total electric bill. Even if the on-site power plant charged just as much as the utility for each kilowatt-hour consumed, every kilowatt generated on-site would save them almost $34. Generate 100 kW on-site, save $3,400. Generate 1 MW on-site, save $34,000. Per month. Every month.

Two points should be immediately clear. First, there is, at least in some areas of the country, a large financial incentive to generate power on-site, owing to the demand charge savings alone. Second, the electric companies in these areas are enjoying all of that nice revenue and will fight the on-site power plant developer every way they can while trying to make the project untenable—hence the appeal of power plant designs like the one in figure 6–1, where there is very little the electric company can do beyond charging extortionate fees for the ABT connection and requiring *Kirk keys* for the generator output breakers. Kirk keys allow control of an electrical distribution system, whereby one circuit breaker cannot be shut without a key that is shared with another breaker—such that the second breaker must be opened before the key can be removed and then used in the first breaker.

To show the exact savings a customer will experience, I model the initial 15-minute interval data as well as the information from the engine-generator set performance model above. In this case, I will use the billing format from the electric utility for Northern California, Pacific Gas and Electric (PG&E); the billing format used by other electric companies around the country will be similar. I reference the E20 rate schedule (available on PG&E's website), which is for large users of electricity (>1 MW).

In the PG&E billing format, the year is divided into two parts, summer and winter, with summer defined as May 1 through October 31 and winter being defined as the rest of the year. There are three billing zones in a day, with all three used during summer but only two used in winter. During summer (PG&E's "Period A"), peak hours are from 12:00 noon until 6:00 p.m., and partial-peak hours are from 8:30 a.m. to 12:00 noon, and again from 6:00 p.m. until 9:30 p.m., Monday through

Friday (except holidays). Off-peak hours are 9:30 p.m. until 8:30 a.m., Monday through Friday and all day Saturday, Sunday, and holidays. During winter (PG&E's "Period B"), only the partial-peak and off-peak zones are used, and their times are the same as during the summer. The only holidays recognized by the utility are New Year's Day, President's Day, Memorial Day, Independence Day, Labor Day, Veterans Day, Thanksgiving, and Christmas Day. There are also three different rate structures that depend on how the power is delivered to the customer. These are as follows:

- Secondary, if the service voltage is below 2,400 volts, typically from the secondary side of a transformer
- Primary, if the customer is served from a single substation at the primary voltage without using a transformer
- Transmission, if the customer is served without the use of a transformer at the higher transmission line voltage

Since we started the genset performance model on January 1, we will begin the financial model there as well. Unless you want your spreadsheet to go further to the right, you may want to put this model on a new tab. I start with the remaining imported power, after the generation assets have done all they could, from the end of the performance model; this constitutes all the power that will still be needed from the utility even though the on-site power plant is running. For simplicity, I address all five of the time zones, and for ease of reading I copy and paste the time of day, day of the week, and date columns next to this *time-of-use* (TOU) zone tracker (table 6–15).

Table 6–15. TOU zone tracker

PG&E TOU New Imported kW					
New Import	**Winter Off-Peak**	**Winter Partial Peak**	**Summer Off-Peak**	**Summer Partial Peak**	**Summer Peak**
130.52	130.52	—	—	—	—

Using color-coded cells within the data columns allows easier tracking of like pieces of data, which may shift from one zone to another over the course of a day. I also do this for the days of the year, to differentiate weekend days and holidays from regular weekdays. Using this method, I arrange the cells such that the "New Import" data are copied to the correct TOU column, for each 15-minute interval, for the entire year. Although this task is painstaking, the trouble is worth it in the end. Once these "new" imported power data are catalogued, I do the same with the original data. I also use the space at the top of the spreadsheet to enter the reference information from the latest edition of the E-20 schedule; since this is subject to change, I always include the date of the data in the table and note whether it was for the primary, secondary, or transmission rate schedule. The maximum is the anytime demand charge, regardless of time of day or day of the week. An example rate schedule is shown in table 6–16.

Table 6–16. Typical rate structure from E20 schedule for January 25, 2017

TOU Zone	Demand Charge ($/kW)	Energy Charge ($/kWh)
Summer:		
Peak	$18.05	$0.14423
Partial peak	$5.01	$0.10738
Off-peak	—	$0.08208
Maximum	$15.67	—
Winter:		
Partial peak	$0.05	$0.10203
Off-peak	—	$0.08832
Maximum	$15.67	—

To create an accurate estimate of the electric bill for each month, you need to get a total of the kilowatt-hours spent in each of the billing time zones, which is the total of all the kilowatt amounts divided by four. Each of the 15-minute interval data kilowatt amounts covers only a quarter of an hour, not a full hour. I start with the months of the year in a column on the left and the quanta of interest along the top of a billing data table. Here I collect the total kilowatt-hours for the winter partial-peak and off-peak zones, as well as the TOU peak kilowatt amounts. For the winter months, that's the partial-peak time zone and the overall, anytime, and maximum kilowatt amount. If the electric company adds a demand charge for winter off-peak billing hours, insert an additional column, as shown in table 6–17.

Table 6–17. Demand charges on new electric bill

Month	Winter Off-Peak	Winter Partial Peak	TOU Demand	Max Demand
January	135,813.43	84,472.7	838	838
February	114,422.53	89,665.29	838	838
March	151,834.97	129,331.46	958	958
April	81,394.62	84,544.21	852	852

I continue down table 6–17, through the rest of the months of the year and across to include all of the summer billing time zones and demand charge peak kilowatt amounts. Once the calendar year is completed, along the right-hand side, it's easy to add the kilowatt-hour amounts together to determine a monthly total of the electricity consumed by the facility. I then copy this data table and paste it a couple rows beneath it, so I can use the reference billing information to calculate the expenses associated with each block. Again, along the right-hand side of the table, it's easy to add up all of the charges to get a monthly total; similarly, along the bottom, it's easy to calculate the totals of each billing time zone and demand charges over the course of the year. From all of this, whatever information is desired can be found; for example, it can be seen how much of the electric bill from any given month was from demand charges and from which time period, which can be helpful when deciding whether to modify the customer's processes to better align with TOU zones.

Averages, as well as total energy costs, are a handy metric for comparisons, such as the average cost per kilowatt-hour for operating the customer's facility during the summer months compared to the winter months. Breaking down each of the billing for each month into the overall average rate ($/kWh) and the demand charge component ($) to get the net electricity rate charge (total $/kWh) can prove the impact of the demand charges on the overall cost of operations.

In addition to knowing what the new electric bill for the customer is predicted to be, it's necessary to know what the do-nothing option would look like. All that is required is to just do to the original 15-minute interval data the same tasks that were done to the new utility import data. Now you can provide a direct comparison, month to month, of the costs associated with using electricity from both the electric utility and the on-site cogeneration power plant—almost. All that's been done is to calculate the net utility charges; the cost to the customer of power from the cogeneration power plant hasn't been identified yet, but that's next.

If the on-site power plant is owned and operated by a company that is independent from the customer, then that entity will need to figure out how much to charge the customer for electricity and thermal energy, and there are several approaches possible. In general, whether the project is done by an *energy services company* (ESCO) or by the owners of the industrial facility, certain expenses must be taken into account. The first is the cost of doing the preliminary investigation of the site and identifying the needs of the end user of the energy produced, including all of the details up front. This cost, mostly in person-hours and travel, is added to the costs associated with the intellectual property of the project; that is, the design phase and all of the research into equipment options that can fulfill the requirements set forth in the systems engineering and project management controls. Second are the costs associated with all of the permitting and construction applications that must be filed with the local government authorities and electric utility, including the installation of any new pipelines for natural gas, water, and other fluids. By far the largest share of the costs is for the procurement of all the miscellaneous pieces of equipment, from engine-generator sets to electrical switchgear to piping, wiring, and paint. Third is the cost of building the project, which, if not managed well, can be very expensive. With all of these EPC costs in the budget, the O&M costs also need to be figured out, from which an end-of-life (total life-cycle) cost figure can be estimated. Once done, efforts can be made to reduce the slope of the line from the origin of the project to its end, as shown in figure 6–5.

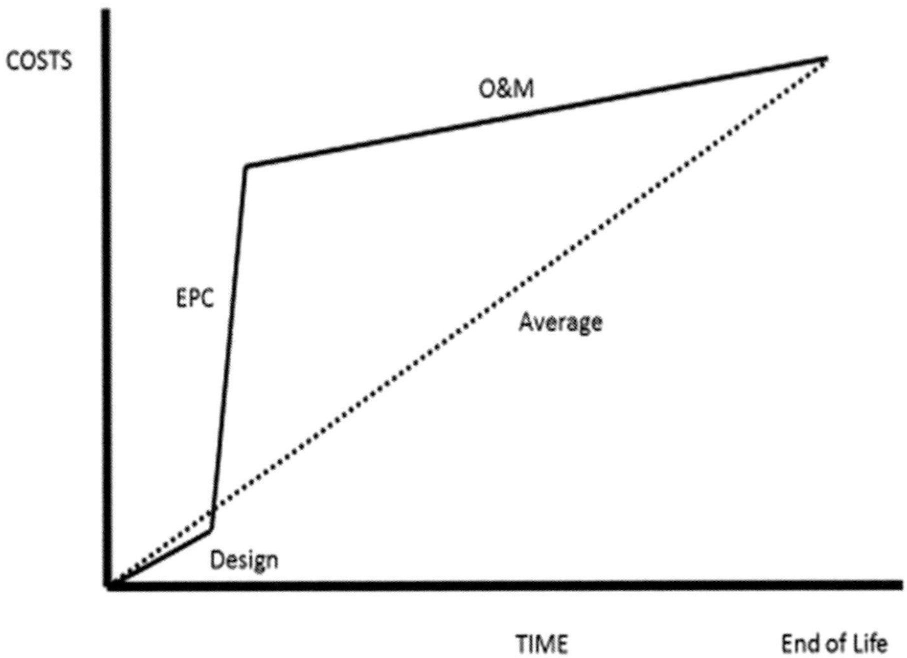

Figure 6–5. Average costs

If the company that owns the commercial or industrial facility wants to also own the on-site power plant, then a profit margin is less of a concern than is a return on the investment; however, in my experience, the facilities that could most benefit from having an on-site cogeneration power plant do not have the $4–$6 million it would take liquid (just sitting around with nothing else to do with it). Additionally, they often are not "power plant people," so they wouldn't really know what to do or how to do it (or it'd already be done), and direct ownership of the power plant therefore becomes unlikely. Still, I have known more than one such company, where the idea of someone else owning and operating a power plant at their facility was simply unacceptable. I know more than a few chief executive officers who would rather go without the benefits of on-site power than go into that much debt and buy their energy at what must be something more expensive than wholesale. In the vast majority of cases, it is actually a win-win situation for everyone, because the customer gets all the benefits of having an on-site cogeneration power plant, not the least of which is a significant savings in energy costs, and the power plant developer (owner-operator) also gets a decent return on investment for their efforts.

Reality versus Wishful Thinking

One of the most important objectives of calculating (or at least estimating) the real costs associated with operating and maintain the cogeneration power plant is having an accurate model that reflects the realities of how the equipment will *actually* function, compared to the nameplate specifications. Failing to insert reality into the cost estimates is the number-one reason why cogeneration power plants perform below expectations or end up as outright failures. This happens all the time on the smaller scale of our everyday lives. How many people have bought a smartphone only to realize the data package they got was more expensive every month than they expected and will be that way for the next two years until the contract ends? How many people have bought a house at the upper end of their budget without knowing how much it will cost them to pay for the air conditioner during the summer and the heater in the winter to keep the place at even a moderate temperature? Many expenses in our everyday lives end up being more costly than anticipated. Fortunately, we can avoid this pitfall with our power plants by being diligent and realistic in our analysis.

The two biggest drivers of the cost of operating and maintaining a cogeneration power plant are the heat rate of the prime mover(s) and the overhauls of that equipment. Other significant costs are the salaries of the power plant operators, the expenses associated with all of the consumables at the plant (oil, spark plugs, etc.) and unanticipated repairs to the equipment. Many of these items, whether direct costs or reductions in revenue-generating performance, are tied to the weather and the percent loading of the equipment, and fortunately for us we've already got that information thanks to the performance model. In the specification data sheets the equipment manufacturers provide, there is information in the form of either a table or graph of the change in the maximum output of the unit versus ambient temperature and altitude. This information can be transferred into a table in a new tab of the spreadsheet model and graphed; from this, the equation for the curves can be extracted.

I start with calculating the maximum power output of the generators and use this as my reference point for each 15-minute interval optimization rather than the nameplate data, which I use to determine when there is sufficient load to start the unit, among other factors (since those operational concerns do not change with changes in the weather). Next, I look at the heat rate, which is the amount of fuel needed to generate 1 kWh of electricity. I typically use a table including the engine efficiency and the thermal efficiency, for 50%, 75%, and 100% engine loads (table 6–18).

Table 6–18. Efficiency distribution

	100%	75%	50%
Engine BHP	566	425	283
Engine efficiency	35	32.9	29
Thermal efficiency	51.5	54.3	58.2
Total efficiency	86.5	87.1	87.2
Fuel consumed (BTU/BHP h)	7,266	7,737	8,780
Fuel used (BTU/h)	4,112,556	3,288,225	2,484,740
kW	375	281.25	187.5
BTU/kWh	10,967	11,691	13,252

Correlating the percent loading on the engines (in this case) and the amount of fuel consumed on a per-kilowatt-hour basis makes it easy to calculate the amount of fuel burned by all three engines in the performance model during each 15-minute interval over the course of a year. This not only gives a much more realistic figure for the actual fuel costs compared to the nameplate data but also is very helpful when doing sensitivity analysis: How do changes in other parts of the power plant affect the amount of fuel consumed (and why)? When using the actual temperature-corrected upper limit on the generator's output instead of the nameplate data, a much more realistic picture of just what the inverter-charger will need to be able to contribute to the main AC bus can be seen.

For the maintenance costs of the engine-generator sets, a few variables should be identified. These will depend on the specific equipment and cover not only what maintenance items are required but also when those items are to be performed. Some *preventative maintenance system* (PMS) items will be due after a set number of run hours on the individual pieces of equipment, whereas others will be calendar-based recommendations. PMS items oil changes are easy to calculate, based on periodicity, quantity of oil, cost of new filters, anticipated downtime, among other factors.

Overhauls, though, are a different matter. If the equipment manufacturers are unwilling to share with you the costs associated with a top-end and/or a complete engine/turbine overhaul, I would be hesitant to purchase anything from them. I have dealt with several manufacturers' sales representatives who, rather than give me a quote (or even an estimate) on the costs associated with an engine overhaul, will instead try to persuade me to buy a maintenance warranty plan. These plans usually come in three flavors (good/better/best, silver/gold/platinum, etc.) and are associated with various amounts of coverage for routine maintenance and parts only, at the low end, up through what amounts to an effective insurance policy covering everything, including the overhauls. The costs for these plans is typically on a dollars-per-run-hour basis, but this does not take into account that some, if not most, of the time the unit will be running at less than full load—in other words, generating less than maximum revenue. Other manufacturers offer plans based on a dollars-per-kilowatt-hour basis, and financial versus performance analysis will need to be done over the life cycle of the power plant to determine which is the better choice. Keep in mind that these maintenance or extended-warranty contracts are typically only for a relatively short duration, usually needing to be renewed or renegotiated immediately after the first complete overhaul.

OEMs also provide information on the temperature and flow rates of the engine's exhaust at 50%, 75%, and 100% load levels. Incorporating this information into the performance model allows for the calculation of the amounts of thermal energy harvested by the heat recovery system. Most of the time, this energy is sold to the customer on a $/BTU basis. Comparing the expected rate of usage of thermal energy by the customer to that harvested by the power plant allows for an estimate of the revenue from thermal energy sales.

7

Cogeneration Power Plants: Case Studies of Successes and Failures

Question: What do you call the person who graduates last in their class from engineering school?

Answer: Engineer.

Even though you can learn a lot by studying successful on-site power plants, the application of that knowledge has its limits. This is because one of the reasons to study other people's successes is to emulate what they did in a future project of your own; however, the main problem with this idea is that each on-site power plant will be unique. A cookie-cutter approach, repeating one basic design, doesn't work well in the world of distributed generation. Each site, each customer's use of energy, each set of circumstances, and potential for growth are too different from each other to allow a "one size fits most" way of addressing their unique problems, needs, and demands.

By contrast, one can also learn a lot by studying the failures of others—and the application of that knowledge has no limits. Granted, making adjustments to your engineering designs based on avoiding the previous mistakes of others will not ensure that your designs will be successful, but at least you will have mitigated one aspect of possible failure. Another difference between the two approaches is that successful designs are remembered for more general aspects, such as minimizing the piping and head loss for a reciprocating engine's onboard water pump. Failures, though, are usually remembered in very specific ways, such as what can happen if check valves are not installed on individual drain lines that lead to a common header.

In emergency situations, you need to know not only what to do but also what not to do. Putting water on a fire to extinguish it is generally a good idea, but not if it's a grease fire. If you were learning about wilderness survival, would you rather learn how a few people survived their ordeal, or how the majority of people could have survived but failed? Knowing the few correct steps to take is essential, but so is knowing the host of things to avoid doing. Ideally, we would all know both sides of the issue; however, each new design is just that: *new*. It has been said that a mistake is simply evidence that someone tried to do something. I submit that in power plant design and engineering, just as in wilderness survival, there are more ways to screw up than there are to get right. In any case, there are more examples of failures in this chapter than there are success stories, and these lessons learned may be applied universally. Appendix A lists common failures and omissions to learn from and anticipate.

Many of the on-site cogeneration power plants built in the past 10 years have failed (or performed suboptimally) because of poor design and engineering of the project at the beginning. In addition to the general engineering of a plant, you need another layer of engineering encompassing equipment maintenance, safety, integration of new equipment into existing systems, and management of the customer's expectations, for example.

Using case studies from existing power plants, I will discuss the following projects:

- A 4.1 MW CHP plant using OEM 1's equipment
- A 2.2 MW CHP plant using OEM 2's equipment
- A second 2.2 MW CHP plant also using OEM 2's equipment
- A 560 kW CHP plant using OEM 3's equipment
- A 4.6 MW CHP plant using OEM 4's equipment
- A 2.3 MW CHP plant using OEM 1's equipment
- A 375 kW CHP plant using OEM 1's equipment

In addition, common CHP project design and engineering mistakes will be discussed.

In a draft of this chapter, I had identified the OEMs by name to better inform the reader of what problems (and benefits) could be associated with different brands of equipment. Unfortunately, in these Litigious States of America, it turns out that I cannot specify the names of the OEMs, even if I were to say complimentary things about their equipment. As it is, most of my comments are not complimentary. This is because the various manufacturers have had mixed success as they tried to reconfigure their diesel-fueled engine designs to burn natural gas and biogas. It's been a steep learning curve, and some have done better than others. Criticism, however slight, grants warrant to be sued these days, and in the interest of financial self-preservation, I must necessarily be vague in some areas. That said, the descriptions below are of real power plants and the real failures they experienced—as well as the successes they enjoyed—and are not fictionalized in any way. In each case, the people involved had the best of intentions and certainly did not set out to fail. Yet they did fail—sometimes spectacularly. Why? How? There are many reasons, some conspiring together like "perfect storms." Others are less obvious. In any case, I'm a big fan of learning from the mistakes made by others, and learning about how these good people ended up with failed power plants may help you to avoid the same fate.

Having been in the power plant industry for over 30 years, one recurring theme run through all but the rarest of circumstances: Those who design power plant systems, almost without exception, have never spent any time actually working in such facilities. In an abstract way, the designs may look great on paper, and the operational theory may sound good too (or at least plausible). The problems come about when that theory meets reality. Sure, the design might have a fluid system valve located 15 feet off the ground; although that works on paper, try operating that valve on a cold winter's night. I am a firm believer in the idea that *anyone* who designs a power plant system should have at least two years of operational experience working in a power plant environment *before* they begin to envision new plant designs. In general, those with more hands-on, real-life operational experience will develop better engineering system designs than those who have only academic experience.

When I give guest lectures on engineering at universities, I give real-world examples from existing power plants, including some of those used as case studies in this chapter. Without exception, I am told by the students that they were never taught those kind of lessons in all their years of school and study. In (almost) all of the institutions, the focus is on the equations associated with fluid dynamics, modeling the thermal fluxes in a heat exchanger, or some other academic exercise. The lessons and classwork projects are theory based, relying on mathematics to give the students the knowledge they need to go forth and become good engineers. Although such exercises serve their purpose, they do not prepare engineering students for the real world.

The takeaway from this is simple. If you are not an engineer but want to benefit from having an on-site cogeneration power plant at your facility, then you should ask your chosen engineering firm certain rather pointed questions. These should include how much experience the design team has in the real world, and not just in a cubicle farm in an engineering firm's office building. Appendixes B and C list such questions, but the answers would be meaningless without an understanding of the consequences associated with the given problem areas—which is why those are described in terms of the project case studies below.

Project 1

This is a food processor in central California, which operates 24/7 for about 350 days a year—taking only the last 2 weeks of each year as scheduled outage time. This is a large, multimillion-dollar-a-year company that makes everything from ice cream toppings to coffee drinks, as well as bottling private-label fruit juices. As such, they have a large electrical load, a large thermal load, and, usually but not always, a high degree of energy demand coincidence.

The facility was using their own 800 HP natural gas-fired boiler to make steam for their processes and imported about 3.6 MW of electricity from the local grid, which made them a good choice for on-site power generation. Furthermore, the local electrical utility power was far less than reliable, dropping out about 15 times a year. Even if the power dropped for only half a second, it was enough to bring everything to a stop as all of their PLC-driven production lines needed to be reset. In these cases, the employees would frantically reset the PLCs, to get the production lines going again as quickly as they could. Some of the production lines could be restored in this way; however, product lines using an aseptic system would have to be cleaned and resterilized before production could continue. Imagine a small-diameter pipe within a pipe. One pipe carries the food material, the other pipe carries either steam, hot water, or chilled water—depending on the food item and whether it needs to be cooked or held at a high or low temperature. If it is of the high-temperature variety, then any stoppage (however short) of the pump(s) pushing the food product through the pipe will cause the product to cook in place. It doesn't have to be stagnant for long before it is ruined and cannot be packaged and sold. For products like baby food, the pumps pushing the food down the piping cannot even slow down, let alone come to a momentary stop, without ruining the entire batch.

Thus, for each power outage (no matter how small), the facility lost about four hours of production, as well as having wasted product. After years of complaining to the local electric utility about unreliable power, the owner of the food processing facility agreed to have a third-party ESCO install and operate an on-site cogeneration power plant. It was promised by the ESCO's sales representative that an on-site power plant would cure all of the power-related ills the facility was experiencing, and the owner-executive of the food facility agreed to the deal. As is often the case, the sales representative was not a power systems engineer; thus, the goal of the sales representative was to close the deal, not ensure that all promises made about the performance of the power plant are accurate or even realistic.

This disconnect between the promises made by any given salesperson and the real abilities of the power plant is one of the main sources of frustration of the typical customer once the facility is online. That many of these promises are not recorded in writing and may have even been misheard (or misunderstood) by the customer is beside the point. The result is that the customer will be disgruntled, believing the deal they agreed to (and the commitments that were made) is not being honored by the ESCO. If not addressed directly by all the affected parties early in the process, misunderstood expectations can fester over time, changing from disappointment to outright distrust.

In this case, the third-party ESCO leased a plot of land at the food processing facility to establish the footprint of the power plant, as part of a 15-year PPA to sell steam and electricity to the customer. The cogeneration power plant included two reciprocating engine-generator sets, each with an HRSG. The heat from the engine coolant was not harvested, but instead went directly to a dump radiator.

Consequently, considerable thermal energy that was available to harvest from the engine's coolant was instead lost; however, no one picked up on that detail. This is an example of the kind of red flag that, in hindsight, is crystal clear; even though, to someone not familiar with power plant systems, everything seems fine, it's *not* fine. The goal is to be as energy efficient as possible, not throw away thermal energy that would cost real money to make.

In this case, the ESCO was new to the game and had recently purchased the project prior to the start of construction from another company, and they were not deep in experienced power plant personnel. It was deemed perfectly acceptable to just use the exhaust of the engines as a thermal source to make steam, and it would be 10 years before they added a system to recover the heat from the engine coolant. As part of the deal, the ESCO purchased the original on-site 800 HP boiler, making 125 psi steam. The steam generated by the HRSGs of the cogeneration power plant would be operated at a slightly lower pressure than the gas-fired boiler and connect to the facility's main steam header via a check valve—this way, the primary source of steam for the processing plant would be the boiler, with the HRSG steam acting as supplemental. This is illustrated in figure 7–1.

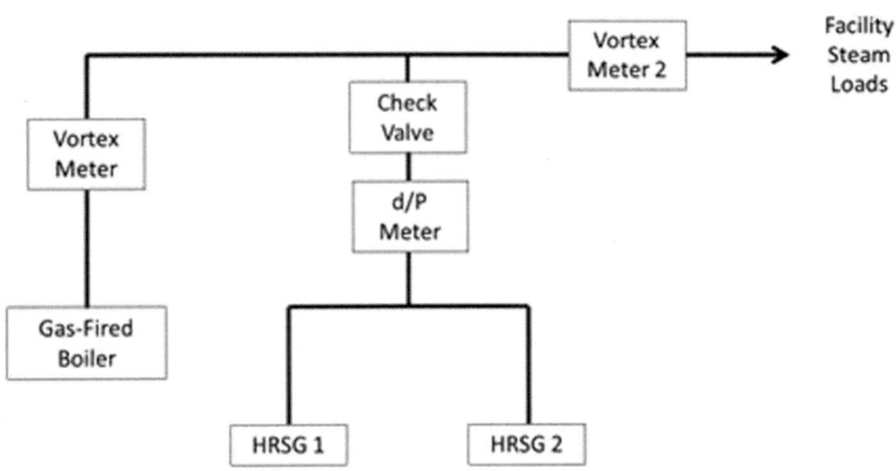

Fig. 7–1. Food processor steam system

Even though this boiler was sold to the ESCO during installation of the cogeneration facility, the food processor facility's staff continued to operate the boiler as if it was still theirs, regardless of any protests by the cogeneration power plant operators. After all, the steam produced by that boiler was the lifeblood of the food processor, and as a critical element to their production, it continued to get a lot of attention. This should immediately bring to mind safety-related concerns, since boilers do represent some risk; nevertheless, complaints to the owner of the facility fell on deaf ears since that boiler was the heart of their operation and critical infrastructure. Consider the liability issues that would arise if one of your customer's personnel were to be injured by improperly operating a piece of your facility's equipment, regardless whether they were trained or had permission to be there.

Even though there was interaction between the ESCO owner-operators of the on-site power plant and the customer prior to the cogeneration facility being built, engineering-based interfaces were not addressed comprehensively. For example, no consideration was given to the chemical treatment needs of the feedwater for the gas-fired boiler and the HRSGs. Specifically, they could not use the same water treatment and distribution systems. Although the original design of the HRSGs was for an industrial facility, food-grade steam has a different set of regulatory requirements and the design corrosion control chemical treatment intended for the HRSGs could not be used; once an acceptable food-grade chemistry for both systems was found, it came at greater operational cost than was anticipated. A more experienced ESCO would know that integrating power plant systems (e.g., steam and heat exchangers) with a food processor requires more specialized equipment (e.g., rated by NSF International) than those in industrial sites.

A rough start

The first nine months of operation of the power plant were not good. An endless stream of mechanical or electrical failures prevented the on-site power plant from performing as advertised to the customer. Distrust between the customer and the ESCO was growing daily. The relationship was getting so acrimonious that the engineering staff of the host facility started to believe that the HRSGs of the cogeneration power plant were not actually supplying any steam to their facility—rather, they were perceived to be acting as condensers, taking precious steam from their gas-fired boiler and away from their processes.

A steam transient was observed, to disprove these (mis)perceptions. It was clear that the gas-fired boiler's pressure started to decrease as steam flow increased and that, once pressure in the main steam header reached the operating pressure of the HRSGs, the rate of lowering system pressure was reduced as HRSG steam began to flow across the check valve. With all three sources of steam flow in service, the pressure transient was much less severe. Once the plant transient was over and the entire steam system recovered, the steam isolation valves of the HRSGs were shut. On the next transient, the difference in steam pressure in the boiler was considerable, and as steam flow increased, the boiler's pressure continued to drop—so fast that the chief engineer of the customer's facility stopped the test. It was now clear to him that the steam flow in the HRSG was a buffer, rather than the main source (across a check valve), but still a vital part of the overall steam supply.

Let's consider this for a moment. With a power plant already constructed and fully operational—and with all of the steam flow and steam pressure gauges ready to examine and the steam piping well known—it still took a physical test of the overall power plant's response to an increase in steam demand for the customer's chief engineer to accept as fact that the HRSGs were really supplying steam. That is the deep level of distrust between the customer and the ESCO to which this contractual arrangement had sunk. But wait, there's more.

Vortex versus differential pressure

Further confusion (and consternation) arose because of the nature of the two steam flowmeters that were used: the customer's was a vortex flowmeter in the main steam header just downstream of the gas-fired boiler (and again in the combined steam header), and the other was an orifice differential pressure (DP) flowmeter belonging to the owner-operators of the cogeneration power plant in the combined header from the HRSGs (see fig. 7–1). Both metering technologies are sound, and each has advantages and disadvantages. If you've ever paddled a canoe and stuck your hand in the water, you may have noticed the swirling vortices in the wake created by your hand. A vortex steam flowmeter works similarly. A small probe is inserted into the middle of the steam flow path, and the steam is forced to go around it, which it does—first to one side, then to the other side, and continuing back and forth in a regular, predictable way. This creates little vortices in the steam flow that can be measured, and the rate of these vortices over time is directly proportional to the amount of steam flowing through the pipe. One problem is that a minimum amount of steam flow is needed to start the process, just as if you placed your hand in the water when the canoe wasn't moving, you wouldn't generate any vortices either.

Vortex steam flowmeters are sized for the range of expected steam flow they are to measure. In this case, the customer's main boiler's steam output was less than 30,000 pounds of steam per hour, but the meter they installed was designed for flows up to 121,000 pounds of steam per hour. Thus, even though the installed meter would work, it would require a considerable steam flow before the vortex action began; consequently, a fair amount of steam flow went down the pipe unrecorded. An analogy for this situation would be if you were to cook a prime rib roast and wanted to use a probe thermometer to monitor the internal temperature of the roast as it cooks. You need a probe with a range of 0–500°F, but since a friend makes you a great deal on a used 0–5,000°F probe, you use it instead of a more precise fit to your needs. Since the roast only needs to get to 130°F or so, it should be clear that, while the probe at hand will work, it won't be as accurate as the better tool for the job. Similarly, you could replace the speedometer in your car with one that has a range of 0–1,600 mph and it would still work just fine—except the difference in the needle's position from 35 to 65 mph would be about the width of the needle itself. While functional, it would make it more of a challenge to know exactly how fast you're going through a school zone. Regardless whether you're measuring the temperature of a roast in the oven or the speed of your car, if the measuring device didn't actually start working until you were above 5% on scale, it wouldn't be very helpful.

A DP steam flowmeter works by many different methods that are all based on creating some resistance to the flow of what is to be measured or changing the diameter of the pipe the substance is flowing through. In this case, a plate (or baffle) is inserted into the pipe that has hole in the middle of it with a very precisely known diameter. This creates a slightly higher pressure on the supply side of the orifice than on the downstream side, and that difference in pressure increases proportionally to the amount of increasing steam flow. It is a comparatively low-technology method of measuring steam flow when compared to a vortex flowmeter, but it works well and, if sized for the expected range of steam flow, can be precise enough to generate billing statements (when selling steam by the pound). Note that when a DP meter is used, there will be some recordable difference in pressure as long as there is at least some amount of steam flowing through the pipe. The linearity and accuracy of that measurement aside, any measured steam flow is going to be more accurate than zero.

So, from the very beginning of the on-site power plant project, the customer's second vortex steam flowmeter—one was for the gas-fired boiler, and the second was in the main steam header downstream of the HRSG steam pipe connection to measure total steam flow—was always reading lower than what the HRSG meter read. That is, the customer subtracted the steam flow from the boiler from the amount recorded by the total steam flowmeter, and this should be a cross-check to confirm the amount of steam coming from the HRSGs. The problem, however, was that, during low-flow periods, the DP meter kept recording steam flow while the vortex meter defaulted to zero measureable flow—hence it always read less than the HRSG meter.

This rancorous atmosphere surrounded the on-site power plant and the ESCO's personnel for the next 15 years. Multiple lawsuits were filed, and it was never possible to regain any of the customer's trust. An old expression states that trust is like a mirror; once it has been shattered, even if it is glued back together, it may work but will never be the same as it was before. Unless you want to work in a continuously toxic environment, I cannot overstress the importance of establishing and maintaining a close relationship with your customers where mutual trust is fostered.

Broken promises

Another big contributor to the customer's animosity toward the ESCO was that the on-site power plant was supposed to fix the problem of the electrical power going out because the installed generator capacity was greater than the customer's total electrical load. In theory, this was supposed to allow shifting to island mode when the local power is out and then shifting back to parallel operations with the utility once it's back. Unfortunately, the control software for the switchgear and the engine control software did not work well together. It took about two years to fix this problem, including installing load-shedding software and testing observed by representatives from the local utility, especially for resynching and closing the utility breaker.

First of a kind

The deal between the ESCO and the OEM to buy the chosen equipment was not made at the local level. That is, the local OEM's distributor was not invited to participate in the discussion. Instead, the national corporate office was the conduit for the deal, along with the parent company of the ESCO, whose headquarters were four states away. In this case, the specific engines in the engine-generator packages were the first of their kind in the western part of the country. It came as a bit of a shock to the local distributor that their mechanics would now be required to work on these units, since they were not consulted in the sale of the equipment. That their mechanics had no experience with these particular engines was also a bit of a concern.

There were two large problems with this arrangement. First, all of the work the local mechanics performed included extra time to master the learning curve while the maintenance team became familiar with the engines. Second, an additional time delay in any maintenance work on the units resulted because of the arrangement for obtaining spare parts. When a part was needed, the maintenance team had to obtain concurrence by the engine experts at the corporate headquarters before the given part(s) could be ordered. Beyond the built-in time delay in getting the parts to the mechanics at the site, this system also induced high levels of frustration among the local maintenance team. The increase in downtime also frustrated the ESCO's power plant operators—and, not surprisingly, the customer as well.

Derating

Another problem that plagued the new engines was that they were subject to an unexpected derating; that is, their output would be below what the nameplate data said it should be for no apparent reason. At this particular location, it was common for the first 80°F day of the year to occur in March, and have daily highs stay above this temperature through October. During the summer, from May 1 through September 30, the high temperature of the day is 100°F or above, with very little or no rain. These high ambient temperatures caused the derating to occur via two methods: The first is the derating effect associated with the temperature of the intake air to the engine (as described in chap. 5); the second effect was due to the sizing of the dump radiators in the engine coolant circuit, because they were sized for average conditions. Unfortunately for the ESCO, their location was not average, so the installed dump radiators could not reject as much heat from the engine's coolant as needed for the bulk of the calendar year. With higher-temperature coolant inside the engine, the control system did what it was supposed to do and began to reduce the load on the engine. All of this was happening during the season when the customer needed the maximum amount of energy and therefore had to import more electricity from the local utility than they expected. This unanticipated extra utility cost definitely did not please the customer.

Another problem this site had that could be considered a form of derating was that it employed the use of a dilution air blower in the engine exhaust streams. The emissions control system catalyst that the design engineer chose was not rated for the temperature of the exhaust from the engines. Thus, it was necessary to reduce the temperature of the engine's exhaust before it got to the catalyst, and this was done by diluting the exhaust with ambient air. Why anyone thought this was a good idea is a mystery to me (unless maybe the design engineer's brother-in-law sold the chosen catalyst). In any case, a blower was used to insert outside air into the exhaust stream, thus lowering the temperature below the limits of the chosen catalyst. It should be clear that this also lowers the available heat for the HRSG; as temperature in the HRSG goes down, steam pressure goes down, and steam flow goes down. This is not ideal when the goal is to maximize thermal efficiency. Also, it introduces a single-point safe component. If the blower goes down for any reason, they can't run the engine-generator set or HRSG. And yes, this happened. Repeatedly.

Catalyst issues

In the physical layout of this power plant (fig. 7–2), the engine-generator sets were on the far left-hand side. The control room and switchgear were in a small enclosure in the middle. The HRSGs and dump radiators were along the right-hand side.

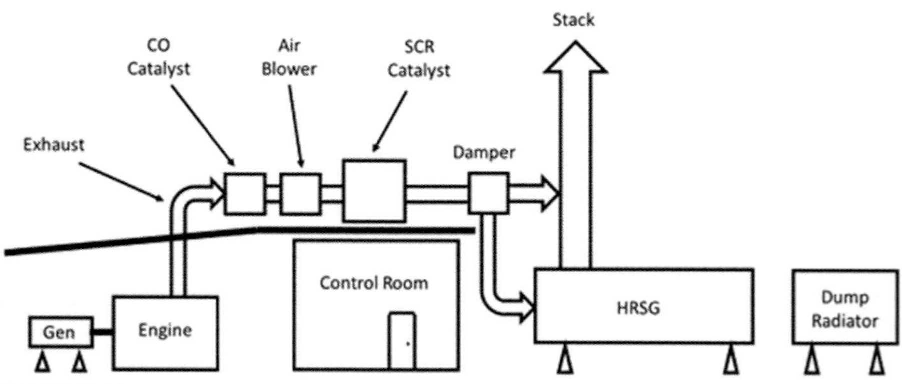

Fig. 7–2. Project 1 configuration

The piping for the exhaust of each engine, coming off the top of each unit, was routed straight up prior to a 90° bend to the right. Here, about 12 feet off the ground were the CO catalyst units, which turned the CO into CO_2 before heading into the dilution blowers. The piping for the now-diluted exhaust then went along the roof of the control room enclosure to where the catalyst unit was for the SCR. This was a catalyst housing designed to hold square blocks of the catalyst material, about 2 feet wide by 2 feet long and 5 inches thick, in a 3 × 3 arrangement. There were provisions to install up to five rows of these 3 × 3 sets of catalyst blocks in the housing; however, because of thermal warping of the housing over time, the blocks became increasingly difficult to remove and replace. This warping also had the negative effect of allowing gaps to form in between the catalyst plates, through which exhaust flow could slip. Furthermore, with the catalyst housing at rooftop height, it was difficult for the maintenance crew to safely access the removable panel on the side of the housing, let alone swap out the catalyst plates.

Multiple rows of catalyst plates were needed because each row was guaranteed by the manufacturer to remove only a certain percentage of the NO_x and SO_x emissions; thus, to meet the standards for the final emissions to the atmosphere, multiple reductions had to occur over these multiple rows. Under nominal operating conditions, three rows of the catalyst plates were sufficient, with the first row doing most of the work. Also, if there was ever a significant problem with an engine, such that it blew carbon soot or other particulate material into the exhaust header, the first row of catalyst plates would act as a filter, shielding the supplemental rows from harm. In theory, since the first row would be exposed to the most amount of chemical exchange, it could be moved to the back row after a year of operational exposure. This would move what had until then been the second row into first place, and this cycle of row transposition would extend the life span of the entire collection of catalyst plates. The experience of the ESCO, however, was that nine new catalyst plates (one row) were required to be purchased each year in order to maintain the plant's emissions within the specifications required on their permit to operate.

This power plant used a tank of urea and an injection system for each engine's exhaust header to provide the necessary chemistry for the catalyst to perform its function. The urea system for each engine included a pump and a control mechanism built into a small steel cabinet located next to each HRSG. These urea pumps had stainless steel discharge piping that plumbed the urea into injection nozzles in the exhaust header upstream of the catalyst housing, where the stream of urea would become a very fine mist as it mixed into the exhaust stream. The ESCO that was operating the facility soon learned that this kind of urea-based system was better suited to a baseload situation, where the system would be operating continuously, rather than a load-following situation with frequent shutdowns. Just as when you shut off the water supply to your garden hose at the spigot, it doesn't mean the water is removed from the entire length of the hose downstream. Similarly, if one of the engines was off-line long enough for the exhaust header to cool down, the liquid urea still in the injection nozzles would then crystalize, thus blocking the small discharge ports of the nozzle. Thus, the next time the engine was started, there would be no flow of urea into the exhaust stream—and, therefore, no cleanup of the emissions in the catalyst bed. This required the power plant operators to climb up, retrieve the injection lances, clean them, and then reinstall them—regularly and frequently. The alternative was to have several spare injection lances standing by, to use on a rotating basis, and have the operators clean the fouled lances when time permitted.

Another problem the ESCO encountered was that since urea is a corrosive agent, the impellors of the pumps would slowly be eaten away. While the system would indicate that the urea pumps were working, there was no urea actually flowing, and once again, the emissions would be affected. The only advantage to using the urea-based system was that any imperfection in the threaded connections of the stainless steel piping used at the facility would develop a growth of white crystals, making them easy to see. After several years of trying to make this urea system work, the ESCO ultimately modified the equipment to allow them to use anhydrous ammonia instead of urea. Using DOT-rated tanks of ammonia also eliminated the need for a urea storage tank, along with its spill and hazmat control protocols.

Dampers

In an effort to allow the operation of an engine with its associated HRSG unavailable, the engineer who designed this power plant included two exhaust flow paths. Just downstream of the catalyst housing was a flow-control damper, which controlled the amount of engine exhaust flow to either the exhaust stack or the HRSG. Unfortunately, this damper was not 100% effective, so even if all of the engine's exhaust was directed up the exhaust stack, there was still some flow to the HRSG. This flow provided enough heat that the HRSG could not be open for maintenance while the engine was running, nor could the HRSG be allowed to be run dry. Not only did this damper fail to do what the design engineer expected it to do, but anyone with experience operating such dampers could have easily predicted as much.

The point this underscores is that there are plenty of power plant system design concepts that look great on paper but don't work well in the real world. Unless the team designing your cogeneration power plant has real-world operational experience, prepare to be disappointed with the final results.

Coolant expansion tanks and dump radiators

In the initial design of this power plant, the thermal energy from the hot engine coolant was not harvested, but rather had to be rejected to the atmosphere via a large dump radiator. The coolant expansion tanks were on a raised platform over

20 feet high, to provide for changes in the volume of the coolant due to temperature changes, as well as to provide a small amount of static head. On these tanks were the sensors and switches to control the high- and low-level alarms, but there was not a simple way to access the coolant. That is, to add to the system or to drain any coolant, a power plant operator would have to climb a ladder to access the tank with a garden hose. Later, additional piping was plumbed so these actions could be done from ground level. The extra cost of this addition was negligible, yet the improvement in operations safety was significant—for example, no one would have to climb a ladder that was sitting on a patch of ice while a 40-knot wind was blowing. Again, if the design engineer had had any operational experience, these kinds of features would have been built in from the beginning.

In this power plant, each engine had its own dump radiator to reject the heat from the engine's coolant. Unfortunately for the ESCO that operated the power plant, there were a couple of problems with these radiators. As described above, the radiators were undersized, leading to operational problems (derating). Ambient air was blown over the coils by a single, large fan with a motor-driven set of blades, instead of by multiple, smaller fans. While some would argue that using a more modular approach to radiator fans is beneficial because it reduces parasitic electrical losses, a larger, negative point must be recognized here: Specifically, when only one motor is used, the radiator fan becomes a single-point safe item. If that motor fails, then the engine cannot be run, and half the power plant becomes useless—all due to the failure of a single part. Also, being a large motor, it was very heavy. This meant that it took several people from the maintenance team to replace these motors, and such evolutions had to follow safety guidelines to prevent an injury if the motor were to fall or even swing on a chain. Using multiple, smaller motors, each with its own fan, eliminates these operational problems while mitigating the safety concerns.

Another problem with the radiators in this project was their physical location: at the edge of the footprint of the power plant and right next to a set of railroad tracks. Several times a day a train would roll by, kicking up dust, plastic bags, and trash that would be blown onto the cooling coils of the radiator. The net effect was that these radiators required far more frequent and intensive cleaning than the power plant operators had anticipated.

Steam header

Each HRSG at this plant was about 8 feet in diameter and about 30 feet long, mounted on supports that allowed access to the underside of each unit for maintenance purposes. The steam header came straight up from the top of the HRSG for about 4 feet before the main steam isolation valve was installed. This meant that, in the event of a steam-line rupture, a power plant operator would have to find a ladder and then climb up on top of the HRSG to gain access to the isolation valve. This means that any industrial accident could last a prolonged time before the danger could be abated. Compounding these safety concerns, the tops of the HRSGs did not have a walkway or handrails, and the steam piping was not insulated; therefore, the poor operator who had to go up there was subject to burns if he stayed and falling if he didn't.

Beyond the basic safety implications of having uninsulated, hot steam piping, there was an operational concern. During the winter, when the power plant was shut down (for any reason) long enough for the piping to equalize to ambient temperature, all of that metal acted as a condenser for any steam introduced during the next power plant startup. In this case, it was not just a few feet of pipe, but rather about 100 feet of exposed metal. If the goal was to operate the power plant at the maximum efficiency possible and in the safest ways possible, then the cost of adding some lagging to the steam piping from the beginning was insignificant.

Air-conditioning

The control room for this power plant was not large, only about the size of a two-car garage. In addition to the human-machine interfaces, it also contained the switchgear for electrical power distribution, including the intertie circuit breaker, an output breaker for each generator, and four distribution breakers (all at 12.47 kV). The control room also contained a couple battery-charging systems (control power), storage for spare parts, and a small worktable. All of the electrical switchgear generated a fair amount of heat, and without any windows or other ventilation, it would get oppressively hot inside. Opening the doors at either end of the room did not help either, because the noise from the running engines was excessive and the outside air temperature was consistently over 100°F (peaking around 115°F) during the six months of summer. The engineer who designed the facility included a small, window air conditioner about the size of a microwave oven. No surprise, this small air conditioner could not keep the temperature in the control room below what felt like the boiling point of plastic, and this took its toll on both the operations crew and the electronics inside the control systems and protection relays.

After several failures of the window air conditioners, a new air-conditioning system was installed outside the enclosure, which provided several tons of cooling capacity. Again, this kind of problem should have been foreseen, but without personal experience in power plant O&M on the part of the engineer designing the plant, it was up to the owners to make such corrections. The problem was that it all looked good on paper before construction began.

Operations-induced problems

The first year of operation for this power plant experienced a perfect storm of problems, and although the people involved in the project had the best of intentions, all of the damaging consequences were ultimately self-inflicted. From the initial personnel assigned by the ESCO that owned the power plant to the way the equipment was operated, an unfortunate sequence of events impeded the success of the project for years.

The history of this particular ESCO helps explain how and why things turned out as they did for this project. The parent company was well established in the construction industry, headquartered in a state in the South. With specialization in projects for the oil and gas industries, petrochemical refineries, and utility-scale power plants, the electricity market in California caught their attention. Three small cogeneration power plants were for sale, fully developed but not yet into construction, so the parent company formed as a new subsidiary an ESCO that would take on the projects. The procurement of the equipment, including the control systems, would be handled by the parent organization through corporate contacts known to the executives, and personnel would be sent to California to manage the construction efforts. Plenty of smart people were involved, from both the parent company and the newly formed ESCO. The problem was that none of them had any significant O&M experience with power plants like those they were buying. For example, they envisioned being able to operate these power plants via remote control from the corporate headquarters, two time zones away. As incredulous as that may sound to those familiar with power plant O&M, the wonders of modern control systems (even as they were back then) and Internet connectivity made it seem realistic.

Since the ESCO was fluent in the construction industry, the building of their first power plant went reasonably well. Unfortunately, not all of the electricians and the other construction personnel were used to dealing with energized equipment, let alone proficient with power plant O&M. Two of the construction electricians agreed to stay on after the plant was built to help commission the facility and then assume the role of power plant operator. Their learning curve was essentially vertical and, predictably, it did not go well.

Recall that these engine-generator sets were new to the West Coast, as were their control systems and the control system for the switchgear. Without knowing any better, the struggling ESCO crew were operating the engines in emergency-generator mode for the first 8 months. That is, instead of operating in load-following mode, where the outputs of the generators matches the changing load demands, both engines would try to stay online—including when loads dropped very low. Instead of one engine shutting down once the electrical load fell below the capacity of a single generator, both units tried to stay online continuously, regardless of load level. As a result, the engines would trip off-line owing to insufficient load when the food processor shut down operations around midnight. Some days, especially on the weekends, the engines would trip off several times. Each time, the power plant operator would receive a message and would return to the facility to restart both engines.

As a consequence, the engines spent a significant amount of time operating at very low load levels, including extended times at 10%–30% load levels, without the benefit of follow-on full-load operations. When a stationary reciprocating internal combustion engine is operated at low load levels, the pressure within each cylinder is much lower than under a normal (relatively high) load, so cylinder components become overlubricated. The piston rings cannot regulate proper oil flow. Carbon buildup occurs, among other problems. The result? Both engines had a lot of head-related maintenance and unexpected downtime. Here are some of the statistics, from a memorandum written about one year after the commissioning of the power plant:

> *For engine #1 (4,527 run hours) and engine #2 (5,264 run hours), the following observations and conclusions can be made. Engine #1 ran for an average of only 77 hours between faults resulting in engine failure, and an average of only 21 run-days between the need for major repairs. This resulted in approximately 32 days of inoperability while awaiting parts and repairs over the past six months. Furthermore, when the engine was operable, it exhibited an unexplained derating of over 20% of engine capacity.*
>
> *Engine #2 ran for an average of only 188 hours between faults resulting in engine failure, and an average of only 18 run-days between the need for major repairs. This resulted in approximately 14 days of inoperability while awaiting parts and repairs over the past six months. The local OEM maintenance technicians had no specific experience with the engines at the facility prior to their attempts to repair and maintain them. The resulting additional time required to analyze engine faults was compounded by the delay in obtaining parts due to the local OEM dealer requiring authorization from OEM corporate HQ before allowing the required parts to be ordered. In addition, service technicians have not always been available, resulting in additional delays to engine repairs.*
>
> *The engine and switchgear control system has never been tuned to shed electrical loads upon a loss of utility power. During seven losses of utility power, the running engine has tripped off-line. The engine and switchgear controls were supposed to shed load until total load was within the capacity of the running engine. All attempts to facilitate the coordination of the local OEM personnel and switchgear personnel have failed. The switchgear manufacturer has said the local genset OEM dealer is responsible for such tuning, and the local OEM technicians say it is beyond their scope.*
>
> *In the course of the last 6 months, there have been 7 failures of the system to automatically go into island mode of operation and 46 days of engine unavailability, resulting in lost revenue, a loss of operational flexibility, and severe customer dissatisfaction.*

Engine control was later switched to a load-following system.

Outcome

Once the engine control system was in a load-following mode, most everything worked much better. Both engines required extensive maintenance on their heads, including some dry ice blasting, but ended up running reasonably well afterward. It took some significant coordination among the equipment vendors to work out some of the control issues, including the ability to shift back and forth from island mode. Eventually, an additional heat recovery system was installed to harvest the heat from the engine coolant loops, thus improving the overall thermal efficiency of the power plant. Customer relations, however, remained less than ideal, ranging from unpleasant to hostile, owing to the long history of unsupported promises made by the ESCO's original sales representative and the long history of operational problems.

Project 2

This project was also for a food processor, though a smaller facility than that of Project 1. This customer also had a shorter operational season, sometimes only 300 days per year, and usually only two shifts of eight hours each per day. The cogeneration power plant used two engine-generator sets, with each generator rated at 1.1 MW at 480 V AC, from a different manufacturer than in Project 1. Again, there was no thermal harvesting of energy from the engine coolant loops, but only from the exhaust. The exhaust from each engine was passed through a CO catalyst unit, converting it to CO_2, before going through a single-pass HRSG. The exhaust discharge from the two HRSGs was combined into a single outlet plenum, where a large monolithic catalyst was installed to clean up the emissions before being sent up a single exhaust stack. The use of a single exhaust stack was supposed to simplify the construction of the power plant, but this had the unintended consequence of making the annual emissions source testing more difficult. That is, not only did each engine have its own individual emissions test at full load, but there was also a third test running both engines simultaneously at full load. The equipment layout is shown in figure 7–3.

The engine coolant's heat was rejected to the earth's atmosphere via a dump radiator that used six separate fans for finely controlled heat exchange and minimal parasitic load. The coolant expansion tanks were raised high off the ground, as in Project 1, to provide static head. As before, there was no way to sample the coolant or add coolant from ground level. This was later fixed by plumbing in piping from the tank down to a manifold more accessible to the power plant operators.

One new problem with the physical arrangement of the equipment was that there was no way to remove any of the tubes from the HRSGs. The clearance between the engine and the HRSG was minimal, only about four feet, and clearance was only a little more than that between the other end of the HRSG and the dump radiator. Thus, if there were ever a problem with a tube in the HRSG (essentially

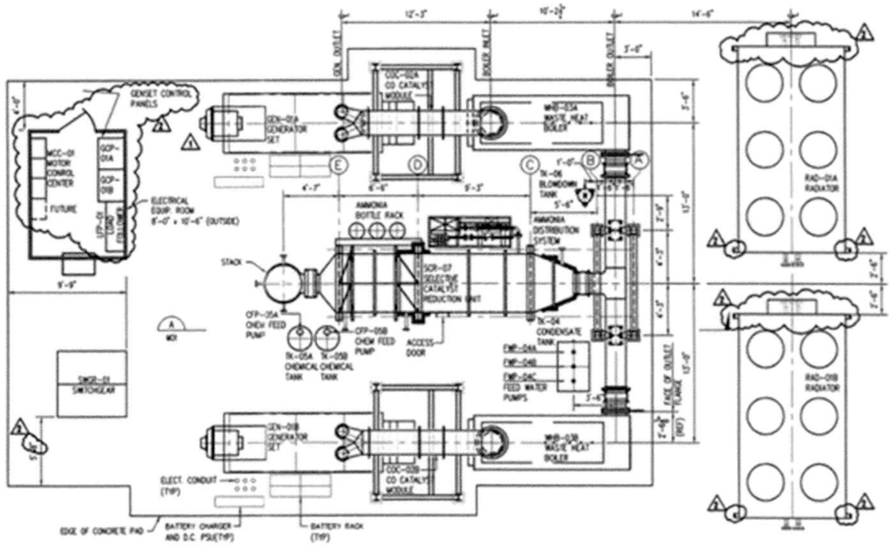

Fig. 7–3. Single-exhaust-plenum equipment design layout

just a large shell and tube heat exchanger), the only recourse possible would be to plug the tube. These HRSG units were really just an unfired, single-pass boiler, but being able to replace tubes should always be a consideration when positioning equipment on the given footprint for the power plant.

The emissions control system used three DOT-rated tanks of anhydrous ammonia, in parallel, as the source of ammonia that was injected into one end of the common exhaust plenum via a series of lances and an air blower. This allowed mixing before the treated exhaust entered the SCR catalyst, and the amount of ammonia flow was proportional to the total kilowatt amount the two generators were supplying to the main AC bus. The catalyst used a vanadium pentoxide compound in its matrix, and whether it was due to poor mixing of the ammonia in the exhaust plenum or the chemistry of the catalyst the result was a nonlinear response. That is, the emissions coming out of the exhaust stack did not change linearly with a linear change in engine-generator load. This resulted in somewhat higher emissions at low load levels and a higher amount of ammonia slip at high load levels (the amount of ammonia not used in the chemical reactions in the catalyst that is released to the atmosphere).

Water where it's not supposed to be

The night before the annual emissions test, the engines would be shut down around sunset so that the exhaust plenum could cool down overnight, with the dilution blower from the ammonia injection system running to flush clean air into the void. Early the next morning, the access panel on the side of the exhaust plenum would be opened, granting access to the interior. The idea was that once the inside air was deemed safe, one or more of the maintenance crew would climb inside the combined exhaust plenum and proceed to vacuum the interior face of the catalyst, which was about four feet by four feet. All of the accumulated dirt and rust was cleaned out of the plenum, making the catalyst system as prepared as possible for the subsequent emissions test. This was the standard operating procedure for the annual testing required by the local air pollution control district. However, on one of those mornings, the crew could not immediately enter the exhaust plenum. On arrival, the supervisor, acting also as the safety monitor, noticed the air dilution blower was not blowing air into the ammonia injection lances. The physical layout is shown in figure 7–4.

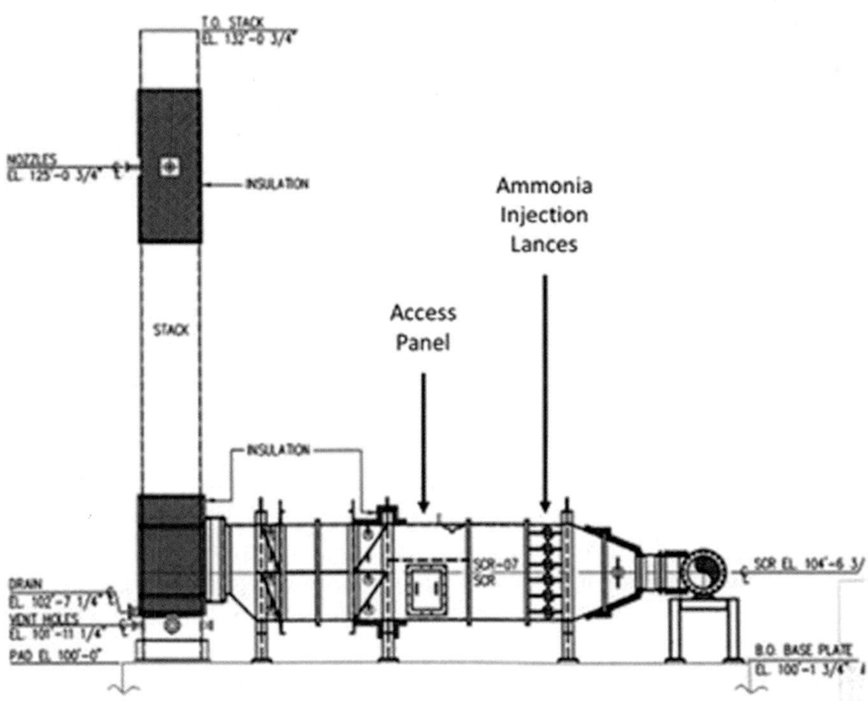

Fig. 7–4. Combined exhaust plenum arrangement

The blower's circuit breaker had tripped, so an investigation began. The discharge from the air dilution blower was coupled to the ammonia distribution system via a rubber sleeve that was about four inches long, and this sleeve had developed a crack. This crack was not so large that air couldn't continue to flow into the ammonia injection lances, but there was another problem: water was leaking from the crack. There had been no rain at this facility for several weeks, so the puddle of water is what caught the attention of the maintenance supervisor. Given that the exhaust temperature of a running engine was about 1,000°F, it was a mystery where this water could be coming from, since it continued to dribble from the rubber piping connection. The maintenance supervisor decided to open the exhaust plenum access panel and, looking inside, found about six inches of standing water inside the plenum. Now the mystery was getting really interesting; how the heck did so much water get into an exhaust header, with no interconnection to any of the fluid systems in the power plant?

The answer turned out to be a surprise, and it was ultimately the result of a combination of factors, not least of which being the assumption that common engineering design considerations had been followed when they had not. As with any boiler, the power plant operators would have to perform periodic surface and bottom blowdowns of each HRSG. These had to be done manually, since there was no automatic system in place to measure the total dissolved solids (or any other water chemistry metric). Unfortunately, the operators were not well trained or experienced in power plant O&M practices, so some would leave the blowdown valves open for shorter lengths of time than would others. Although there was a water treatment chemical mixture that was pumped into the HRSGs along with the softened, deaerated feedwater, the local water supply had very high amounts of calcium and manganese in it. As the water in the HRSG was boiled away into steam, the concentration of these impurities created a white sludge in the bottom of each HRSG. Anyone who has lived in places with very hard water will be familiar with this sludge.

During a blowdown, the sludge buildup was passed into some piping to the quench tank before ultimately going into piping that led to the facility's sewer connection. Unfortunately, these impurities acted the same way a cholesterol plaque does within an artery in the human body, and over time, the interior diameter of the discharge piping became more and more constricted, to the point that a six-inch diameter pipe had only about one inch of flow space. The power plant operators, not realizing this, would continue to open the bottom blowdown valves to discharge the sludge from the bottom of each HRSG. The problem was that there were other drain lines from around the power plant that fed into the common discharge header, and one of those was from the base of the exhaust stack (to drain away any collected rainfall). Ultimately, when the blowdown from the HRSG met the flow restriction in the discharge pipe, backflow of that water along the drain piping from the exhaust stack filled the stack with HRSG water. This water then flowed through the catalyst unit of the SCR into the exhaust plenum, filling it with several hundred gallons. The water level got high enough inside the plenum to enter the ammonia injection lances and thus lead to backflow into the air dilution blower, where it could be discovered thanks to a small crack in the rubber connector between the blower and the ammonia injection system.

In any liquid discharge system that has several points of drainage into a common header, one classic design element is a check valve installed in each of the supply lines. The purpose of the check valve is to prevent back-feeding of liquid like this power plant experienced. Such design elements are so standard and common sense that they were assumed to be in the system by the ESCO who was operating the power plant. However, such assumptions can be very costly, when not verified. In this case, the water poisoned the bottom 8–12 inches of catalyst material in the SCR unit, accelerating the need for its replacement, at about $40,000 to the owner of the cogeneration facility. In the end, blank flanges and/or check valves were installed in the discharge system, and the piping from the HRSG to the sewer line was replaced.

Electrical distribution problems

The electricity supplied to the customer by the local electrical utility came from two separate feeds to the two halves of the facility: the north feed and the south feed. These two power feeds were from poles on the street but were powered from two different substations. Part of the reasoning for this was that if one substation were to go off-line for whatever reason, then the customer's facility could open that side's intertie circuit breaker and crosstie power from the remaining utility feed. This also meant that, when installing an on-site power plant, the electric utility required precautions against both utility feeds' ever operating in parallel with each other (connected). Each feed from the utility to the facility had its own transformer and utility breaker, as shown in figure 7–5.

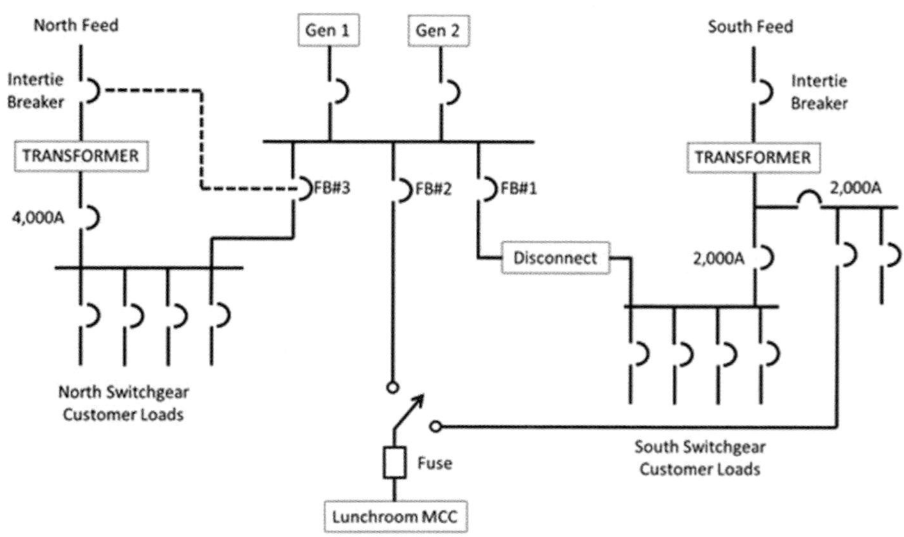

Fig. 7–5. Project 1 electrical distribution

The on-site power plant that was built by the ESCO had two engine-driven generators, each able to produce 480 V AC and up to 1,375 kilovolt-amperes of power. Each generator had an output breaker (protected by a generator protection system [GPS] relay) feeding a 4,000 A main (generator) AC bus. From this generator bus, there were three feeder circuit breakers to supply power to the customer's facility. Feeder breaker 3 (rated at 1,600 A) was Kirk-keyed to the isolation circuit breaker for the north feed from the utility. That is, a key was needed to shut each breaker, and there was only one key; for one of these circuit breakers to be shut, the other breaker would have to be opened first, then the key removed from it so it could be inserted into the desired breaker, which could then be shut. This guaranteed that before either of these circuit breakers could be shut, it would be a dead-bus shift, ensuring that there would be no way for both of these breakers to be shut at the same time. The other two, generator bus feeder breakers 1 and 2, were limited to feeding the customer's loads associated with the south feed from the utility.

Generator bus feeder breaker 2 was rated at 2,000 A and supplied a transfer switch, also rated at 2,000 A. This transfer switch allowed either feeder breaker 2 or the upper bus of the south switchgear to supply power to a *motor control center* (MCC) that was located in the lunchroom of the customer's facility via a 700 A fuse. The actual electrical loads powered by the lunchroom MCC were variable, as equipment would come and go; however, because the 700 A fuse feeding the MCC kept blowing, the customer replaced it with an 800 A fuse. Similarly, when the transfer switch was positioned to supply power to the lunchroom MCC via the upper bus of the south switchgear, the supply breaker there was rated to only 700 A and, not coincidentally, it too would trip. It was always a question of which would happen first—the 700 A breaker tripping or the 700 A fuse blowing. Ultimately, when the customer decided to upgrade the fuse to 800 A, the upper bus supply breaker was also upgraded to an 800 A model.

Feeder breaker 1 from the generator bus supplied a *visible disconnect* via a set of 2,000 A fuses. This disconnect was a requirement imposed on the on-site power plant developer by the local electrical utility as a safety device (that could be locked open by the utility). Ostensibly, this was as a way of ensuring there would be no way for the cogeneration power plant to electrify the local grid via the south feed if the utility ever needed to perform work on that section of the power grid. This disconnect also acted as the isolation breaker between the cogeneration power plant's main AC bus and the lower bus of the south switchgear of the customer.

The unfortunate reality of this electrical distribution system was that, even if the transfer switch for the lunchroom MCC was aligned to feeder breaker 2 (and feeder breaker 2 was shut) and everything powered by that MCC was running, the amount of power flowing through feeder breaker 2 was still very low. Compared to the total generation capacity of the power plant, the vast majority of it would want to go through feeder breaker 1 to the south switchgear. Keep in mind that the load-following control system for the on-site generators was looking at the amount of power being imported from the utility through the south intertie circuit breaker (via the IPS relays). The on-site generator control system only registered how many generators were available and how much demand there was to feed the south switchgear—not the proportional distribution through the available feeder breakers off the main AC bus. The problem was that feeder breaker 1 was only a 2,000 A circuit breaker, and the electrical supply and distribution system wanted to put a lot more than that through it to support the south switchgear.

Not surprisingly, feeder breaker 1 would trip on overcurrent, a protection built in to the breaker. If both of the on-site generators were running at the time, this had the effect of dumping almost all of the electrical load, causing both generators to trip off-line via the GPS relay owing to the transient. The customer would now have no power to the lunchroom MCC, but the loads off the south switchgear would not be affected because the utility would instantly supply the power that had been provided by the on-site generators. If the ESCO's crew were not immediately available, the customer's personnel would reposition the lunchroom MCC transfer switch to source power from the south switchgear, thus restoring power to the lunchroom MCC loads. Ultimately, this was both an inconvenience to the customer and a point of annoyance between the chief executive officer and the ESCO.

It can never be proven, but it was suspected that the original operators of the on-site power plant made unauthorized adjustments to the settings of the overcurrent trip for feeder breaker 1. This suspicion came about because, instead of feeder breaker 1 opening on overcurrent, the 2,000 A fuses in the line between feeder breaker 1 and the south switchgear began to blow. Unfortunately for the crew of the ESCO, these fuses are not especially easy to find. As a result, the on-site cogeneration power plant would be off-line for a lot longer than necessary, much to the displeasure of the customer, while replacement fuses were procured and electricians called to the site.

In any case, the problem of what to do with the amount of electricity the on-site generators could produce remained. One solution was to shift the Kirk key from the north-side utility breaker to feeder breaker 3, allowing the on-site power plant to provide power to the customer's loads on the north side as well as to the south side via feeder breaker 1. Doing so and aligning the lunchroom MCC transfer switch back to feeder breaker 2 would then allow the on-site generators to run at full load without putting too much current through any of the three feeder breakers.

Unfortunately for both the customer and the ESCO, there were a couple of problems with this idea. The switchgear on the north side was rated at 4,000 A and was supplied by the utility via a 4,000 A circuit breaker. The supply to this switchgear from the on-site power plant was through feeder breaker 3, but that breaker was rated only at 1,600 A. Only during the off-season were the loads on the north-side switchgear low enough to enable support via a 1,600A supply breaker. During the rest of the year, especially during the busy season, there was no way for the on-site power plant to meet the needs of the north switchgear via feeder breaker 1, and the Kirk key would remain in the utility breaker on the north side.

Once again, the ESCO was left trying to power as much of the south switchgear as possible via feeder breakers 1 and 2. Given the low load that could ever go through feeder breaker 2, this meant a de facto limit on the total generation capacity of the two gensets to that of feeder breaker 3.

This kind of failure is not just the result of insufficient coordination between the ESCO and their customer from the very beginning of the project. Clearly, whoever was designing the electrical distribution system did not understand how the customer's electrical demands changed over the course of a year. Furthermore, although the ESCO could run only one of the engine-generator sets during the main operating season of the customer (splitting the generated power between feeder breakers 1 and 2), this would not be in the interest of the ESCO, whose revenue stream is based on generating as much electricity and steam as possible, for as long as possible during the year. Such problems, ultimately due to a lack of coordination and/or understanding among the parties involved in the planning, design, construction, and O&M of such on-site cogeneration power plants, are a lot more common than most people would expect.

The problem of what to do with all of the electrical power the generators could produce also had an impact on emissions testing. The testing of the emissions from the power plant by the local air-quality board came in three parts: engine 1, engine 2, and then both engines simultaneously. For the duration of the 30-minute test, each engine was supposed to be running at full load. This was fine for the engines running one at a time, but impossible for the combined test; there simply was nowhere to put the electricity that would be generated. A solution was found in the use of a load bank. This artificial load, like a giant toaster, would be connected to the electrical distribution system a day or two prior to the testing. This could be done only when de-energized, so the power plant would have to be shut down—and, once again, any shutdown was annoying to the customer, regardless of the need to do so. Using the load bank allowed the power plant to operate at high loads long enough to prove the effectiveness of the catalyst system's ability to clean the exhaust to within acceptable levels. After several years of operating this way for annual testing, the air-quality board stopped allowing the use of load banks. Instead, they allowed the engines driving the generators to be run at whatever load level there was, regardless of how low that load was. This made the testing a bit more stringent, because the engines weren't burning the fuel as cleanly at the lower operating efficiency levels; however, it did resolve the electrical distribution issue.

Project 3

This power plant project was for another food processor and was almost identical to the power plant in Project 2. It also used two engine-generator sets, each with an HRSG, and a combined exhaust plenum with ammonia injection and an SCR. Nevertheless, there are key differences between the Project 2 power plant and this one, especially in the designs of the physical and electrical distribution systems. In this case, the customer did not want the on-site power plant to be located close to the existing buildings, so the footprint of land that was leased by the ESCO developing the project was located along the edge of the customer's property line, well over a hundred yards away. This resulted in problems with the operation of the power plant, although these unintended consequences should have been anticipated by the engineer who designed the project. The general layout of the facility is shown in figure 7–6.

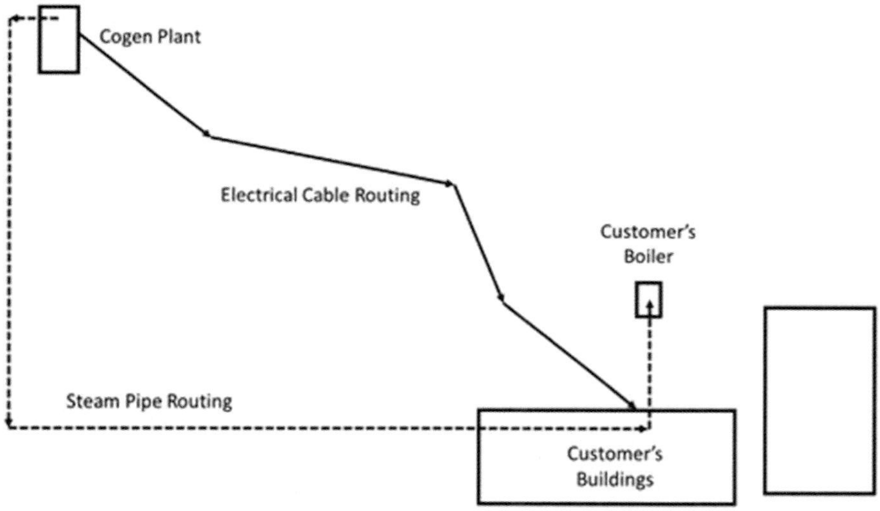

Fig. 7–6. Project 3 physical layout

One important consideration when building a power plant that is remote from the end users of the energy that will be generated is the cost associated with the physical logistics. That is, the power plant will need access to water and, on the other end, access to a sewer or other drainage collection system. Although these piping arrangements are not complex, they will add to the construction costs. Other considerations include whether to route the electrical cables aboveground or set in the ground via a system of conduits. If the site is industrial in nature or will have heavy vehicles driving around, it can be safer to use buried conduits rather than have the power lines high enough above the ground to avoid being hit. Fire codes may require a minimum height standard, allowing fire trucks access to all parts of an industrial or commercial facility. Similarly, transporting the steam and/or hot water generated by the power plant to the customer's facility will necessitate routing the piping around the customer's site logistics.

Electrical distribution problems

In this case, it was decided to bury the electrical cables between the power plant and the customer's building, but there were other buried systems already in the ground (although many were no longer in use). This made it necessary to take an indirect route, with more bends and elbows in the conduit than an ideal case would have, for a total distance of over 400 feet. Pulling the cable through this much conduit, with all of the restrictions, proved challenging during construction. Ultimately, a large forklift was used to pull the cable, because the immense force that was needed to haul such a large electrical cable through that many bends in that length of conduit. If tension on the cable was monitored during this installation, it was neglected, because a couple years after the power plant was commissioned there was a cable failure.

The postmortem analysis of the failed cable run included lessons applicable to all such buried-cable situations. The most apparent is to minimize the deviations from a straight run of conduit, because the increase in friction from every bend and curve will increase the degree of difficulty in pulling the electrical cables through the conduit. Similarly, another consideration would be to use a cable lubricant to minimize the friction-induced physical resistance to getting the cable through the conduit. There are many different kinds of cable lubricant available, and best selection will be based on the individual circumstances of each project; nevertheless, the options bear careful consideration. Also, careful monitoring of the cable tension will mitigate the chances that the insulation on the cable is being stretched enough to cause a thinning of the material or create any other similar weak spot. Another consideration is heat dissipation, in that the cables will get hot from the amount of current flowing through them, and the conduit can effectively act like an oven cooking the cable's insulation; the longer the conduit is, the more intense this oven effect will be.

The cables for this power plant suffered a phase-to-phase short, causing the distribution circuit breakers to trip open. After more than a week, the faulted cables were pulled out of the conduit and examined. The location of the fault was at about the halfway point along the length of the conduit, and the cable from this area was still quite warm to the touch. There was clear evidence of stretched cable insulation in many locations along the length of the cable. The ESCO that developed the power plant did not have any personnel monitoring the original cable installation, instead relying on the electrical contractor. Any seasoned construction manager will understand the importance of quality-assurance and quality-control protocols at the construction site, including the Russian proverb invoked by President Reagan: Trust, but verify. Another cliché comes to mind: Why is there never enough time (or money) to do it right, but there's always enough time (and money) to do it over?

Part of the problem was the amount of current the electrical cable carried, especially the peak amount. This is because, as in project 2, the electrical distribution design was flawed. In this case, there were also two electrical feeds from the local utility, identified as east and west, but they were supplied by the same substation. Both of the on-site 1.1 MW engine-driven generators were connected to the main AC bus via their own isolation circuit breaker, each controlled by a set of GPS relays. There were two distribution feeder breakers between this generator bus and the customer's facility. Feeder breaker 1 connected to the west switchgear via a visible disconnect (as required by the local utility). Feeder breaker 2 was Kirk-keyed to the utility intertie circuit breaker supplying power to the east switchgear, so that the east side was powered by either the utility or the on-site power plant, but never both at the same time. Both the east and the west switchgear were rated at 4,000 A and effectively split the customer's electrical loads between them. This electrical arrangement is shown in figure 7–7.

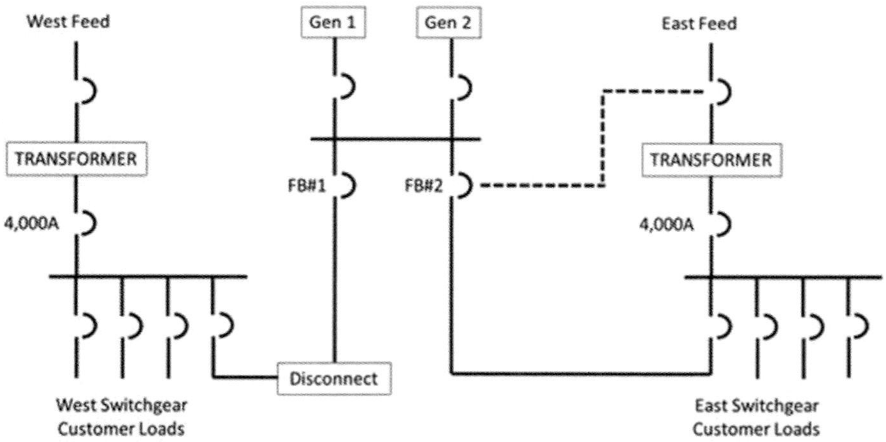

Fig. 7–7. Electrical arrangement for project 3

Shifting the Kirk key from the east utility intertie circuit breaker, as before, requires that breaker to first be opened. Doing so removed all power from the customer's loads supplied by the east switchgear, which was inconvenient to the customer. The key could then be inserted into the on-site power plant's feeder breaker 2, and then that breaker could be shut, restoring power to the east switchgear. Even if both of the on-site generators were off-line, power would be routed from the west utility feed to the west switchgear, back-feeding the main AC bus of the on-site power plant via feeder breaker 1, then through feeder breaker 2 to the east side. The limiting choke points in this arrangement, electrically speaking, were the two feeder breakers in the power plant's switchgear and the visible disconnect, all of which were rated to 2,000 A. With this limit in place, the customer could not power all of the loads on the east switchgear that were necessary for the operations of the facility during the main season. During some portions of the off-season, this 2,000 A limit was not restrictive, but it was still an inconvenience to shut everything down, open the utility breaker, make the key switch, and reenergize the east switchgear. Therefore, this was usually done around 4:00 a.m., to minimize the inconvenience to the customer.

The problem for the ESCO became one of reliability, because any problem that resulted in the loss of power to the east switchgear would result in the customer taking the Kirk key and resupplying power via the east-side utility breaker. With every such occurrence, the customer would become more hesitant to allow the ESCO to have the Kirk key back, thereby limiting the on-site power plant to supplying power to the customer via feeder breaker 1 only. Once again, as in project 2, the ESCO was faced with having to run only one of the two engine-generator sets or running both units at less than full load. However, neither of those options would give the ESCO the return on investment they desired, since revenue was proportional to the amount of electricity and steam the power plant produced. This was also less than optimal for the customer, since the savings against paying the local electric utility for power was less than advertised—even though it was the customer's choice to take power from the local utility for the east side of their facility. The only net advantage was that the power plant now had a built-in spare engine-generator set, so the preventative maintenance on the equipment would not interfere with running at least one genset. That was cold comfort to the ESCO after spending all the extra money on two gensets and hundreds of feet of electrical cable.

Steam piping (condenser)

Just as the electrical distribution required long-distance cable runs, so did the steam system. While the electrical cable took the path of the hypotenuse, the steam piping took the path of the other two legs of the triangle. There was a safety isolation valve at the top of each HRSG, and the downstream steam piping then met to form a common steam header. Shortly thereafter there was a DP-based orifice steam flowmeter for the ESCO, followed by a back-pressure regulator and steam vent. This system allowed a place for the steam to go when the customer was not using any of it and prevented the pressure in the HRSGs from getting high enough to lift the overpressure protection safety relief valves. The common steam header was routed to the nearby fence line at the edge of the customer's property, and then along that fence toward the customer's facility. Once at the edge of the facility's building, the steam piping was routed up and over the road that the delivery trucks used along the back of the processing plant. From there, it was routed to the customer's natural gas-fired boiler, where it joined the boiler's discharge steam header via an isolation valve and check valve. This is shown in figure 7–8.

Fig. 7–8. Steam system distribution

Even though this steam header was insulated, having over 400 feet of piping poses a problem during the colder months of the year. In effect, that length of steam piping became a giant condenser, and some (if not most) of the steam generated by the HRSGs became hot water before it could reach the customer's steam header. Extra steam traps were installed, to capture this condensate from the steam header and remove it from the pipe to minimize the hot water delivered to the customer's steam system. While this did indeed remove most of the condensate, the discharge of these steam traps was deemed to be too far away to bother capturing that hot water and returning it to the HRSG system. Instead, the hot condensate was left to drain onto the ground, made visible during the winter as steam plumes. To compensate for the reduction in steam flow at the point of system interconnection, the ESCO raised the operating pressure of the HRSGs. The idea was to increase the pressure at the source, thus driving more steam flow. As a result, there were three big problems for the ESCO.

The first problem was that the customer was unhappy. The commodity the customer was paying for was steam from the cogeneration power plant, so that the customer's boiler would, in theory, have to produce less of it. The difference in this requirement from before to after installation of the cogeneration power plant, was nowhere near what the customer expected, indicating to them that the ESCO had not delivered on their promise.

The second problem was that the customer noticed that the steam flowmeter used by the ESCO (to create invoices for the amount of steam sent to the customer) was upstream of not only the back-pressure relief vent but also the condensate traps along the steam header. The customer could literally look across his property and see the steam being vented to the atmosphere and condensate spilling onto the ground, all of which he was paying for.

The third problem for the ESCO was that as the weather warmed up, the customer's engineering staff would give the order to shut the isolation valve in the steam header at the boiler, sealing off the steam from the cogeneration power plant. This was done because, now that there was less steam loss due to condensation, the higher delivered steam pressure was lifting the safety valves on the customer's boiler. This act of isolating the steam produced by the cogeneration power plant had several consequences:

- The first was safety related, because now the back-pressure relief valve had to discharge almost 100% of the steam produced in the HRSGs. Failure of this system would cause the pressure in the HRSGs to rise, leaving the individual HRSG overpressure relief valves as the only means of preventing a catastrophe. Some of the steam produced by the HRSGs could still travel down the steam header, which now functioned as a condenser.
- Second, the ESCO had lost one of its revenue streams. This led to discussions with the customer about contractual obligations per the agreed PPA with the ESCO, whereby the customer was obligated to take steam from the ESCO. The customer, however, argued that the delivered steam was inferior in quality and insufficient in volume compared to what was promised during the initial contractual negotiations.
- All of this mutual dissatisfaction and frustration led to less-than-ideal business relationships between the customer and the ESCO.

Had the ESCO been able to convince the customer to allow the on-site cogeneration power plant to be built adjacent to the existing food processing facility, some of these problems would never had happened. The electrical distribution design, however, was still fundamentally flawed.

One of the remedies made to improve the situation was for the ESCO to move the revenue steam flowmeter to a point downstream of the pressure regulating steam vent. This is shown in figure 7–9.

Now it was the ESCO who could observe all of the steam being vented to the atmosphere by the pressure regulator, instead of going down the pipe to the customer, where it would generate revenue. This provided a clear indication of just how much (or how little) steam the customer was actually using.

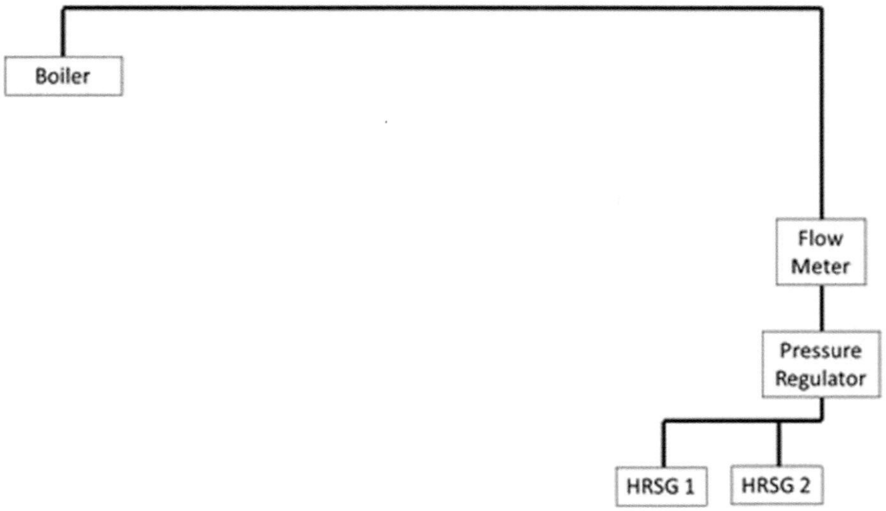

Fig. 7–9. Revised steam system distribution

Financial ramifications

One of the largest expenses for a power plant is the price of the fuel used by the prime mover. The metric of the equipment's heat rate, the number of BTUs consumed to generate 1 kWh of electricity, is only one part of the story. While there are places where one can buy natural gas from suppliers other than the local gas utility, the utilities are usually the ones who own (and therefore control access to) the pipes that the natural gas flows through. As such, there is fee, commonly called a *transportation charge*, associated with pulling gas out of the distribution pipelines. In California, regulations grant a facility that generates electricity a discount on the transportation charge if a sufficient amount of thermal energy is also generated. This lower rate, at about $0.10 per therm of gas used, represents a significant savings compared to the normal transportation charges of about $0.24–$0.39 per therm of gas used (depending on monthly consumption). For a cogeneration power plant using on the order of two million therms of gas a year, the financial benefit of being on the lower, fixed transportation charge is clear.

This discount comes with strings attached, however. On an annual basis, the cogeneration power plants in California must meet a minimum thermal energy standard, in terms of the percentage of the total energy output from the facility. A second threshold, one that is a little more difficult to meet, is the efficiency standard. This compares the electricity and harvestable thermal energies used on-site to the energy of the total amount of fuel used in the process. The key here is that it depends on the amount of thermal energy *used*, not the amount *generated*. For this particular project, every time the customer's personnel would shut the steam isolation valve, the amount of steam used would be less than the amount of steam generated. If sufficient thermal energy was not consumed, the transportation charge would revert to the full price. In this case, the PPA between the customer and the ESCO directly addressed the costs associated with the purchase of natural gas for the power plant. This was essentially a pass-through cost. If the price of a therm of natural gas increased, then the cost to the customer for 1 kWh of electricity and 1 pound of steam also increased proportionally. This gave the customer an incentive to coordinate with the ESCO to purchase the natural gas in long-term strips, locking in the price with the broker selling the gas. By use of this method, the customer would know for several years at a time exactly what would be the cost of each kilowatt-hour of electricity and pound of steam generated on-site. This made the customer's energy budget very easy to calculate with precision.

Problems Common to Projects 2 and 3

Projects 2 and 3 were essentially sister facilities, sharing a common hardware system design and equipment selection. Similarly, they shared common points of failure, especially in the engine-generator sets.

Early in the operation of these power plants, about a year after the warranty period expired, one of the power plant's 1.1 MW generators suffered an arc fault from rotor to stator, causing catastrophic damage to the generator. No definitive cause for the failure was found, so the ESCO owner-operator of the power plant replaced the unit with another of the same type. This replacement took weeks to accomplish because the OEM had to get a new generator from Europe, critically limiting power production during a peak season.

Within a few months, a second generator failed by the exact same mechanism. This time, more in-depth postmortem analysis was done, and it was found that vibration-induced wearing of the insulation in the windings of the generator led to a short circuit, which initiated the arc from stator to rotor. Further investigation by the ESCO found that other companies using this same equipment around the country had suffered the same failure. This analysis was used as leverage on the OEM, and the generator was replaced at a significantly discounted price.

Over the next couple of years, the other generators (at both sites) also failed by the same mechanism. The OEM then provided replacement generators that were more robust, with a heavier frame. The OEM also demanded that the ESCO use a much tighter group of protection set points on the GPS relays, as part of the offset for the discounted replacement costs. These new, more restrictive relay set points caused multiple nuisance trips of the generator output breakers, causing operational issues for the ESCO and an increasing annoyance on the part of the customer. Over time, these more robust replacement generators also failed, again by the same mechanism—for a total of nine failed generators. All of these repeated failures and the many weeks of lost production time while the replacements were ordered and installed led to a severe reduction in the revenue stream for the ESCO, as well as ever-increasing levels of dissatisfaction among the customers, the ESCO, and the equipment manufacturer.

There were problems with the engines on the mechanical side as well. All four of the engines were identical, and not coincidentally, those units are no longer available for sale by the OEM. Early on, valve and other head-related issues arose that were likely due to the original operation of these engines at low load levels for longer periods of time than recommended. There were other issues, though, that are not typical of young engines—that is, with less than 20,000 run hours of operational history. For example, during a routine engine examination that included measuring valve clearances, the engine mechanic performing the work noted unusual wear on the rocker arms. According to the manufacturer's maintenance schedule, these rocker arms were not due to be examined until the overhaul at 30,000 run hours. However, in this case, the engines had less than 12,000 run hours. Furthermore, the wear was excessive enough that the mechanic could not recommend running the engine until they were replaced. Not surprisingly, a check of the other three engines found the exact same accelerated wear pattern on all of the rocker arms.

At about the same time, it was discovered that there were small leaks in the aftercoolers of two of these engines, requiring early replacement. As with the problematic rocker arms, it was not recommended to run the engines until the aftercoolers could be replaced. The net result of all of these mechanical problems is that the ESCO could not operate the cogeneration power plants without risking even more severe, perhaps catastrophic, damage to the equipment. The alternative was to wait for the replacement parts to arrive and repairs to be made. Unfortunately for everyone, these replacement parts had to be made at factories located outside the United States. This added a significant amount of time to the delays already in place, including getting through customs.

In the end, the ESCO was out several hundred thousand dollars' worth of unanticipated equipment maintenance and downtime—for each of the four engine-generator sets. On top of that came the loss of revenue for the ESCO, as well as increased energy costs to the original customer, whose frustration at having a dysfunctional on-site cogeneration power plant continued to grow. Between the failures of the generators and the engines, the overall uptime was far less than expected by everyone involved in the project. From the equipment failures to the design failures, power generation was never optimized.

Project 4

The on-site cogeneration power plant for this industrial facility was limited to producing electricity and hot water and used one RICE with a heat recovery system for both engine coolant and exhaust. The customer stored large amounts of a cryogenic fluid at about –35°F in tanks. To get this fluid into the small tanks mounted on the trailers of large trucks, the facility needed to warm the fluid up to about 25°F and used shell and tube heat exchangers to do so. The heat source for these units was originally water from a river that flowed next to the site. However, due to the inevitable leakage of the cryogenic fluid, causing localized contamination of the river water, a substitute was sought. This ended up being a 250,000-gallon tank of water that was heated up to 80°F by the thermal energy harvested from the cogeneration power plant. At the same time, the electricity generated on-site would go toward minimizing the amount of power that would have to be purchased from the local electrical utility.

The original design was to use an engine-generator set rated at 360 kW and a series of heat exchangers to transfer energy to the thermal reservoir. When this design was presented to the customer, it was determined that the amount of harvestable thermal energy was insufficient for their needs. The design engineer's solution was to replace the 360 kW unit with a 560 kW unit and a pair of immersion heating elements in the water storage tank, thus delivering more heat to the customer. This design was accepted, and the ESCO and the customer agreed to a PPA arrangement. Construction and commissioning soon followed, but it wasn't long before everyone involved started to wonder if the power plant had been built on cursed ground, owing to all the unanticipated problems that kept on compounding. However, even while these problems were unfortunate, they were all contingencies that should have been anticipated.

Application issues

On-site cogeneration power plants are not always a good fit, even for customers that have decent electrical and thermal loads at their facility. Tomato processors, for example, are ideal for the application of an on-site cogeneration power plant, but only for about 110 days a year. The facility for Project 4 was similar in that they had a sustained thermal demand for only a four- to six-week period in the spring of each year and a two- to three-week period in the fall. The rest of the year, demand for their product was sporadic and uncertain, but would be often enough to warrant continuing to operate the facility. In short, this was not the ideal circumstance to use an on-site power plant that was owned and operated by an outside ESCO under a PPA arrangement. Continuous, steady electrical and thermal load, 24 hours a day, for at least 300 days a year, is a much more ideal energy demand profile.

The electrical distribution for this project was pretty simple (fig. 7–10). Electricity was supplied by the local utility at 12.47 kV, which was dropped to 480 V AC by a transformer. The output of the transformer supplied four load cables: two went to a pair of 480 kW immersion heaters (i.e., one cable for each heater) in the thermal storage tank; one went to a 3,000 A supply circuit breaker to the customer's switchgear; and one went to the intertie breaker for the cogeneration power plant. The immersion heaters were installed as a backup source of thermal energy in case the cogeneration power plant ever went off-line, because of a fault or for maintenance, since there was only the one engine.

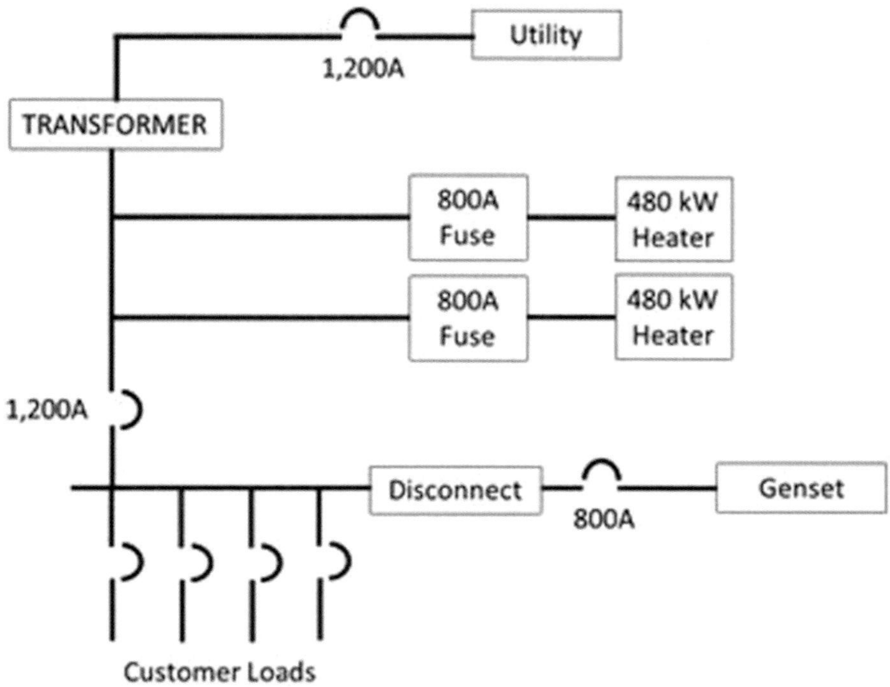

Fig. 7–10. Project 4 electrical distribution

As a practical matter, these heaters were deemed too expensive to use, since the customer had to pay not only for the electricity from the local utility but also demand charges. These heaters had multiple heating elements in them that were individually fused, so an effort was made by the customer to trim the electrical load by removing a few of those fuses. Getting the costs down resulted in insufficient heat to the thermal reservoir, and the customer was unable to find a compromise between the amount of heat desired and the costs associated with getting that amount of heat via the electric heaters. After a couple years of trial and error, the customer installed two natural gas-fired hot water heaters to do the job, since the cost of the natural gas was significantly lower than the cost of the equivalent electric heat.

Without the electric heaters, the overall electric load of the customer was now much lower than the power plant designer had counted on, and it was not going to increase. As a result, the average load on the 560 kW on-site generator was much lower than anticipated. Not surprisingly, this resulted in the engine operating at low to very low load levels for extended periods of time. This made frequent and extensive mechanical repairs to the engine necessary, with associated downtime and customer (and ESCO) frustrations. Furthermore, even when the engine was more than 50% loaded and running well, the thermal harvesting system never worked as well as described by the design engineer.

Heat recovery

The heat recovery system looked fairly elegant as a design on paper, but was never able to deliver on its promise. Once again, without the benefit of having actual power plant operational experience, the design engineer could not anticipate the problems associated with the actual functioning of the heat recovery system. For this plant, hot coolant came out of the engine and went to two heat exchangers, in parallel, in the exhaust path. Exiting these heat exchangers, the now very hot coolant was recombined and put through a plate-and-frame heat exchanger with the customer's process water, thus transferring heat to the thermal reservoir. The engine coolant would then go to a dump radiator (with a temperature-controlled fan) via parallel paths before returning to the engine. All of this engine coolant flow was driven by the engine's internal water pump. This arrangement is illustrated in figure 7–11.

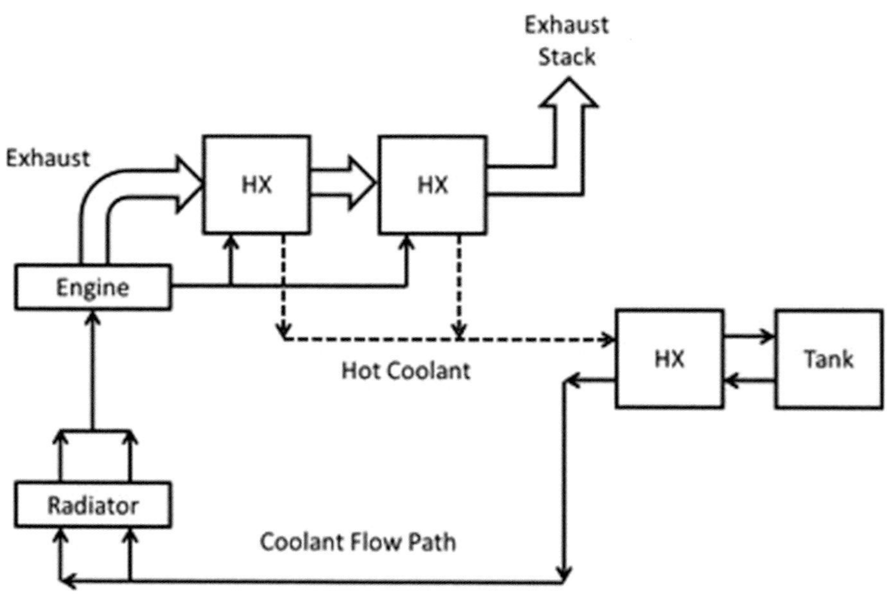

Fig. 7–11. Thermal energy harvesting arrangement

The customer's thermal reservoir was limited to 80°F, to prevent the cryogenic fluid from flashing to vapor in the shell and tube heat exchangers. In reality, the temperature of the water in the tank was most often in the 40°F–55°F range, owing to the thermal load of filling the tanker trucks with the cryogenic fluid. When the cogeneration power plant was off-line, the temperature across the process water plate-and-frame heat exchanger would equalize. Once the engine was started, this slug of cold water would quickly make its way to the engine. The manufacturers of this reciprocating internal combustion engine were smart enough to anticipate this set of circumstances, so they included seven internal thermostats to recirculate coolant flow internal to the engine. This would allow time for the cold coolant to warm up a bit before being exposed to the hot internals of the engine to minimize the amount of thermal shock sustained by the engine block. What this also meant was that, as the engine's coolant was recirculating inside the engine, there was less coolant to flow out of the engine.

As a result of this reduction in coolant flow outside the engine, the coolant in the heat exchangers in the exhaust path spent more time exposed to the heat from the exhaust, thus absorbing more energy. Sometimes this coolant would flash into steam, increasing the back-pressure to the coolant trying to flow out of the engine. Because of this reduction in flow, more heat could be extracted by the considerably colder process water in the process heat exchanger too, allowing the whole process to repeat itself as flow tried to happen through the entire system. Eventually there would be enough thermal energy transfer to get closer to a stable fluid flow (after the internal thermostats were satisfied). Any perturbation to any part of the thermal recovery system would invite chaos. In the end, the thermal recovery system would never perform as the design engineer intended, because the primary driver was the engine's internal water pump, as well as the head loss associated with the external piping and the action of the engine's thermostats.

Containerized

The sales teams that peddle engine-generator sets are fans of selling entire packages that are inside a container. These are similar in size and shape to the containers used for international shipping. Those intended to contain engine-generator sets have been modified to include access doors and ventilation windows, interior lighting, and a variety of other accessories. These enclosed units may work well in cold climates, but I have yet to see a successful installation of such a containerized genset. Even if the equipment sales representatives give assurances that there will be plenty of ventilation, this is simply not the case in reality. These containers become ovens. There is simply too much heat radiated from the equipment for the meager ventilation system to reject, and this results in problems.

I have seen cases where the hoses and wiring harnesses became embrittled, to the point the insulation on the wires cracks and shatters at the slightest movement. When changing a spark plug requires a whole new wiring harness, engine downtime becomes a lot longer than originally thought. The paint on the equipment can blister, and the plastic covers on light fixtures can melt. The solid-state electronic components in the voltage and current regulators on the generator can degrade, requiring frequent replacement. To get sufficient air flow and cooling to the genset, some power plant operators will open the access doors on each side of the container, and although this does indeed resolve the problem of heat buildup, it creates new problems.

The selling points for using a containerized genset include that the power plant will have a lower noise profile. Opening the access doors effectively eliminates any noise containment, so in peopled locations such as a hospital, this can be a big problem. Another selling point for using a container is the supposed protection of the genset from a dusty or dirty environment, as you would find at a feed mill. The reality is that the access doors will still need to be opened periodically for a physical inspection and routine maintenance, and dust and dirt will still get inside. Air, fluid, and oil system leaks do occur, and the longer the equipment operates under leaky conditions, the worse the situation will be.

Containerized systems look nice when new and are a convenient way to ship entire genset packages from the OEM to the end user. Insufficient ventilation, limitations on equipment access (think overhauls), high ambient temperatures, embrittlement of wiring insulation and hoses, lengthy downtime, and increased maintenance costs are all common when a containerized genset is installed in a real power plant. Unless the power plant is sited where the highest temperature of the year is less than 75°F, it doesn't matter who the manufacturer of the equipment is—they will all perform suboptimally.

Safety concerns

One of the issues with this particular site's containerized genset was that the only emergency-stop switches were located inside the container. The emergency stops cause the engine to immediately shut down, prompting the generator's output circuit breaker to trip open. This kind of emergency-shutdown feature is a standard safety precaution. For example, the next time you fill the gas tank of your car, take a look around and you'll see at least one emergency-shutoff switch that would stop all of the gas pumps at once. The concept of having such an emergency-shutoff switch at least a short distance away from the gas pumps is that, particularly in the event of a fire, it might not be possible to reach the gas pump itself. This is analogous to having a fire extinguisher next to the fireplace in your home; it may be impossible to get to it when you need to. Safety items should be placed where they can be accessible under the conditions in which they would be needed.

At one point, a fire did happen inside the genset enclosure, and the plumes of smoke coming from the container gave clear evidence that something bad was happening inside. However, the power plant operators, the personnel of the ESCO, were not at the scene at the time. Neither were the customer's personnel interested in opening the doors of the container and braving the flames to reach one of the emergency-stop switches, so the fire continued. After all, the engine was fueled by natural gas, and for all anyone knew, there could be an explosion at any moment. Eventually, the access doors were opened, the emergency shutdown was activated, and the fire department resolved the emergency. Compounding this crisis, the emergency wrench that was supposed to be tethered to the isolation valve in the natural gas pipeline to the engine was missing; without a handwheel, there was no way for anyone to quickly shut off the gas flow to the engine.

The simple truth is that bad things can happen at any power plant, cogeneration or otherwise, and precautions should be taken to address those potential emergency situations that can be predicted. Fires, spills (including hazardous materials), arc faults, battery problems, and injuries to personnel, for example, can be anticipated and prevention/mitigation efforts need to be a part of the design of the power plant from the beginning. Once constructed, initial and recurring training must be done to ensure that the power plant's O&M personnel, as well as any of the customer's personnel who could be in the vicinity of the power plant, know how to respond to such emergencies. In this case, as part of the repairs following the fire in the genset enclosure, the ESCO installed several emergency-stop switches, at locations all around the power plant, at distances far enough away to allow for safe operation of those switches if another emergency occurred.

Project 5

This is an example of a successful installation, in contrast to the failed systems and/or engineering designs above, and is the replacement power plant for Project 4, after it finally experienced a catastrophic failure. During the planning for the project to repower the site, the entire situation was reevaluated along with the collection of lessons learned from the previous failures.

Almost everything from the original power plant was replaced, although some of the equipment was refurbished or reused. First, the new engine-generator set was from a different manufacturer than the first one and was sized much more appropriately, at 375 kW; this fit the customer's electric load profile much better than the 560 kW unit had, so problems associated with long-term low-load conditions were eliminated. Second, the engine's coolant loop was made as short as possible, to minimize the burden on the engine's onboard water pump. Third, the genset was not enclosed by a container, but simply covered with a roof for rain and sun (i.e., ultraviolet [UV] light) protection, preventing the genset components from being cooked inside a de facto oven.

For both the main coolant loop and the aftercooler/oil-cooler coolant loop, the outlet from the engine was routed to a three-way valve, providing a bypass flow path. The position of these valves was controlled by the temperature of the coolant flowing back into the engine. If the temperature of the returning coolant became too cold, the control system would open the bypass valve, diverting hot coolant flow through piping that bypasses the heat exchanger before reconnecting with the piping that returned coolant from the heat exchanger to the loop's dump radiator. This combined effort, on the part of the loop's heat exchanger and dump radiator, allowed for the rejection of all (or none) of the heat from the engine, as necessary. This coordinated control system allowed optimization by transferring as much heat as possible from these coolant loops to the customer's process water system while still providing protection for the engine. This design is illustrated in figure 7–12.

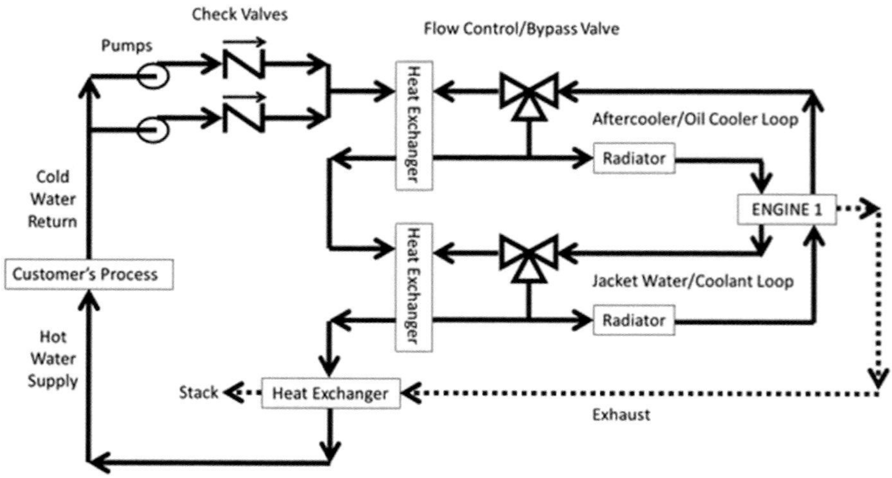

Fig. 7–12. New thermal harvesting design

For the customer's process water, the return from the storage tank was driven by one of two pumps, in parallel, for 100% redundancy. The discharge from these pumps went directly to the heat exchanger for the aftercooler/oil-cooler coolant loop. Because of the lower inlet temperature requirement at the engine and because the customer's process water had an upper temperature limit of 85°F, this arrangement provided satisfactory protection for the engine. The outlet of this first heat exchanger was plumbed directly into the next one, which was for the main engine coolant loop, and the discharge of this unit was plumbed to the two heat exchangers in the engine's exhaust path before heading back to the thermal reservoir. In this way, the customer's process water could easily pick up over 30,000 BTUs per minute under nominal conditions.

The engine-generator set was mounted on a concrete slab, with plenty of clearance between the top of the unit and the bottom of the roof. This provided the necessary room to access the unit for all the maintenance requirements, including overhauls, as well as infinite ventilation. Some covering is certainly desired, not just for rain protection but also to limit the amount of UV-induced embrittlement of the hoses and the insulation on the wiring harnesses. Concrete berms were built in for containment of any spills, whether oil or coolant, with separate drain collections to minimize the chances of drainage or stormwater runoff contamination.

Once installed and commissioned, this power plant worked very well. The engine ran from oil change to oil change, with very few unanticipated shutdowns in its first three years of operation. Life for the ESCO and the customer was much happier than with the first iteration of this cogeneration power plant. It all came down to taking the time and effort to get the project right in the first place. Verifying the customer's needs, using the experience of O&M personnel to influence the design, and going with a reliable equipment vendor all make a significant difference in the performance of the facility.

Project 6

This facility is such a great example of so many things not to do when planning an on-site cogeneration power plant for a prospective customer that it's hard to know where to begin. From the choices of which equipment to use, to the sizing of the HRSG, to the protections against the physical conditions of the site, to the structuring of the PPA—all were mistakes of such magnitude that the power plant shut down within three years of operation. The ESCO involved with the project went out of business soon after this power plant was shut down, as so many others have done. None of them got into the cogeneration power plant business to lose money, let alone fail, but poor planning of the projects from the beginning and a poor grasp of what it takes to actually operate one of these facilities conspired to bring their businesses to an end.

The customer for this project did not need any hot water but did need a fair amount of steam. It was an industrial site, where they made a product resembling a plywood board out of glue and various sizes of wood fibers—better known as fiberboard. Large trucks continuously brought in loads of scrap lumber and other wood products, which were then pulverized to various sizes of dusty bits and sorted into piles. Sawdust was everywhere, blown about by the wind, and was completely unavoidable. Because the electric load was fairly consistent when the facility was operating and the thermal demand was concurrent, the ESCO owning the power plant decided to use a combustion turbine and a large HRSG. Steam demands were anticipated to be beyond the ability of the thermal recovery system on its own, so the ESCO had duct burners installed inside the HRSG to provide additional heat (and therefore create additional steam). The combustion turbine would generate about 4.6 MW of electricity, and the turbine's exhaust would be able to generate about 26,000 pounds per hour of steam flow at 125 psi.

This electrical generation capacity was above the full load of the customer, but it was reasoned that the ESCO would allow for projected growth on the part of the customer. The ESCO accepted the idea of operating the combustion turbine at less than full load—and, therefore, at lower efficiency levels—for a given number of years until the customer expanded their operations. So, from the start, the power plant would be operating at a lower profit margin than it was capable of. Even though I do not have an MBA, I can tell that this move on the part of the ESCO doesn't make much sense unless there were definite plans already in the works for the customer's expansion. Without that, this amounts to nothing more than wishful thinking, and such dreams rarely come true.

Similarly, the duct burners would be able to provide enough extra heat to bring the total HRSG output to about 40,000 pounds per hour of steam flow, all at about 125 psi. At the time of the construction of the power plant, the customer did not need more than 30,000 pounds of steam flow per hour. Thus, wishful thinking was not limited to the revenue from the sale of electricity, but also applied to the sale of steam, compared to the maximum capacity of the equipment.

To address the dusty conditions of the site, the ESCO decided to put the containerized combustion turbine inside a large building. Combustion turbines require significant amounts of air—if the exhaust flow is on the order of 150,000 pounds per hour, think about how much air needs to go into the front end of the unit. This should immediately bring to mind a couple of points:

- The air needs to be relatively clean
- If the unit is inside a building, then that building needs sufficient ventilation to prevent drawing a vacuum

Even though it is true that combustion turbines will always have an effective filter system on inlet air, the operating environment will have a direct impact on how fast those filters will get clogged such that the differential pressure across the filter becomes too high. Many filter systems have a supply of high-pressure air on the turbine side of the filter that is used to create a puff of air backflow, blowing all the collected dust and debris off the face of the air filter and restoring it to nominal operation. If these air filters are exposed to the elements, though, then this back-pressure method can be of limited use; once the dust and dirt get rained on, it can set up like adobe or cement, and mechanical cleaning of the air filter becomes necessary. If the turbine is inside a building, ostensibly to provide a more controlled environment, then the ventilation systems used in that structure must be given serious thought. In addition to the amount of air that must enter the building to prevent a vacuum being drawn, the environment outside the building must be taken into consideration; if it is situated in a dusty or industrial environment, then whatever filters are on the building will need attention.

This facility used a moderate amount of power and steam, so an on-site cogeneration power plant seemed to be a good fit. One problem they encountered was that sawdust was everywhere, inescapable—such that the building housing the turbine never had sufficient ventilation and always had a slight vacuum when the turbine was operating (making it a challenge to open the outward-swinging doors to the building). Furthermore, every time one of the doors to the building was opened, the vacuum would suck in large amounts of sawdust, which would ultimately get trapped by the air filter on the turbine. The problem was that when the turbine's air filter system would try to clean the filter with a puff of air backflow, the dust and dirt would fall to the floor of the building and, if not removed immediately, would just get sucked right back onto the filter again, prompting this process to repeat.

In the typical turbine-based cogeneration power plant layout, the control of the amount of heat going into the HRSG—and, thus, control of steam pressure—is achieved by controlling the amount of exhaust that is directed up an exhaust stack prior to reaching the HRSG. The actuators for this flow damper can be a very weak point in the overall design. Unless the nominal pressure-temperature band for operating the HRSG is relatively large, the damper positioning control system can end up hunting, cycling far more than originally anticipated (see also chap. 5). The physical arrangement of the equipment is shown in figure 7–13.

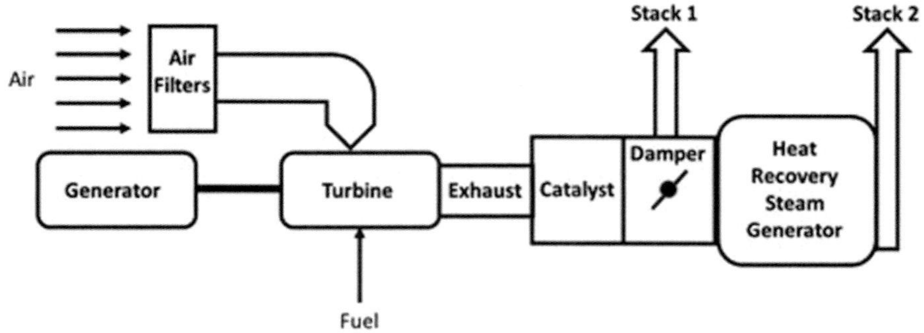

Fig. 7–13. Combustion turbine—based cogeneration power plant layout

With these control systems, whether to regulate the temperature of your home or the temperature (and hence the pressure) within an HRSG, it should be clear that the smaller the control system deadband is, the more the system will cycle on and off. With a turbine exhaust flow damper, subtle changes in its position will result in subtle changes in the amount of heat that is allowed to pass into the HRSG, and if the deadband is tight, that damper's actuator will be cycling a lot. I have seen a 4.6 MW turbine-based power plant brought to its knees by the failure of a $500 pneumatic actuator for just such a flow damper—repeatedly over a single year—because the root cause of the failure was never addressed.

Duct burners

The exhaust of a large combustion turbine will be relatively high in excess oxygen, on the order of 15% or so. This property allows insertion of a postturbine combustion process to increase the thermal energy used to generate steam in the HRSG. This is most often done using *duct burners,* which are nothing more than lances injecting fuel (usually natural gas) and a spark directly into the exhaust plenum of the turbine. One big problem with this, especially in California and other states with tighter emissions regulations, is that in this setup, combustion takes place downstream of the emissions control systems. Thus, the raw exhaust of the duct burners, together with the cleansed exhaust of the turbine, goes straight out the exhaust stack to the atmosphere. Granted, a second SCR unit could be installed on the second exhaust stack, but that would add considerable costs to the construction of the power plant, as well as the costs of maintenance and testing.

Unfortunately for the ESCO in this case, the long-term PPA contract that was negotiated with the customer was done by a salesperson who was not a cogeneration power plant engineer, so there were a few problems with it. First, the agreed sales price for the purchase of electricity generated by the cogeneration system was very low—much lower than the cost of that same electricity from the local electric utility. This provided a strong economic incentive for the customer to include contractual language to ensure that the cogeneration power plant would supply electricity to the customer at all times when the power plant was operational. That is, the on-site power plant would be required to run if it was physically able to do so, regardless whether it made financial or operational sense to do so. For example, if the customer's electrical load was only 1 MW, the ESCO would still be obligated to operate the combustion turbine at that very low load level (for a 4.6 MW unit), even though the heat rate (fuel usage) of the turbine was terribly expensive. Therefore, not only was the customer getting a great deal on the price it was paying the ESCO for the electricity, but the generation of that electricity was also costing the ESCO significantly more money in fuel costs than had been budgeted.

The second big mistake for the ESCO in the PPA contract was the negotiated price for the steam that would be sold to the customer. The salesperson who represented the ESCO believed that, although the ESCO would lose a bit of money on the electrical sales, they would make a killing on the steam sales by charging the customer a much higher price per pound of steam delivered than the customer would pay to make that same steam themselves. The power plant was purported to be profitable by commanding this high price for the steam, plus foreseeing demand for steam being up to 40,000 pounds per hour, and it was believed that the profit from selling steam would amply make up for the loss on the sale of electricity. Unfortunately, this particular salesperson's belief that the customer would use high amounts of steam was a fantasy based on projected steam demand after a future, extensive expansion of the customer's facility. In reality, the customer in this case was never going to use more than about 26,000 pounds of steam per hour, with the occasional surge to about 30,000 pounds per hour. This meant that the steam sales was never going to be high enough to make the power plant profitable for the ESCO. This, though, was not the only steam-related problem.

The duct burners that were installed in the HRSG, for the purpose of boosting the amount of steam that could be generated, were downstream of the emissions cleanup systems. When emissions testing was performed, as required by the local air pollution control district, it was immediately clear that the untreated emissions from the duct burners were well over the limits allowed by the local air-quality board. Without installing additional catalyst, the air-quality board exercised its legal authority and required the ESCO to lock shut the isolation valves in the natural gas fuel supply to the duct burners. Then, after purging the fuel supply lines, the ESCO was required to remove a spool piece in the piping between the isolation valve and the HRSG, thus ensuring the complete inability to ever operate the duct burners again. All of the extra money that had been invested in the extralarge HRSG and the duct burners was for naught.

Operations and management problems

All of these operational problems arose after the power plant was commissioned, but they were not the start of the financial troubles for this ESCO, which had already had trouble managing the construction of the power plant. The original construction manager ended up installing a much more robust concrete pad for the turbine enclosure building than was necessary, which presented a major cost overrun. There were other problems in addition to the concrete pad, and the construction manager was replaced by a more experienced person about halfway through. In the second half of the construction period, efforts were focused on trying to save as much money as possible, wherever possible, to reduce the amount of budget overrun. If corners could be cut without introducing undue burdens on the O&M of the facility, the incentive was there to do so. Why is there never enough time (or money) to do it right the first time, but there's always enough time (and money) to do it over? As you might guess, there were some bugs to work out during the initial testing and commissioning of the power plant. Moreover, troubleshooting these problems was problematic because the wiring diagrams were not entirely useful—in reality, some of the wires were simply not going where they were supposed to. This added delays to the final commissioning of the power plant, which added to the financial burden of being over budget.

Being unable to operate profitably, the ESCO declared the project a failure and terminated the PPA contract with the customer after less than three years of operation. All of the failures associated with this project could, and should, have been anticipated. The planning of a new cogeneration power plant is so often filled with wishful thinking and failures to grasp obvious financial impacts on the O&M of the facility that, in hindsight, are crystal clear and intuitively obvious to the most casual observer. This way of doing things recalls the saying Measure with a micrometer, mark with chalk, cut with axe. The end result will never match the specifications called for. The systems engineering approach (as described in chap. 2) is a proven methodology to mitigate such problems.

Project 7

This is another success story. After over a dozen years of frustration, the ESCO and the customer from Project 2 came to an agreement to repower the site. Because of the massive infrastructure changes needed, it took years of negotiation to agree on which company would pay for which improvement(s) and agree to the terms of a new PPA. Also, the new design was a bit larger than the previous power plant, but the customer was reluctant to surrender more of his facility's parking lot to accommodate the new cogeneration power plant. In the end, the entire electrical distribution system was renovated into a design that allowed the plant to provide the bulk (if not all) of the facility's electrical power needs using one (or both) of the new engine-generator sets all year long. This was done by making the central, primary switchgear a 21 kV system and using two transformers to drop the voltage down to 480 V for the supplies to the customer's north and south switchgear. The engine-driven generators would be making 480 V, and a transformer was used to increase that to the 21 kV of the primary switchgear. This is illustrated in figure 7–14.

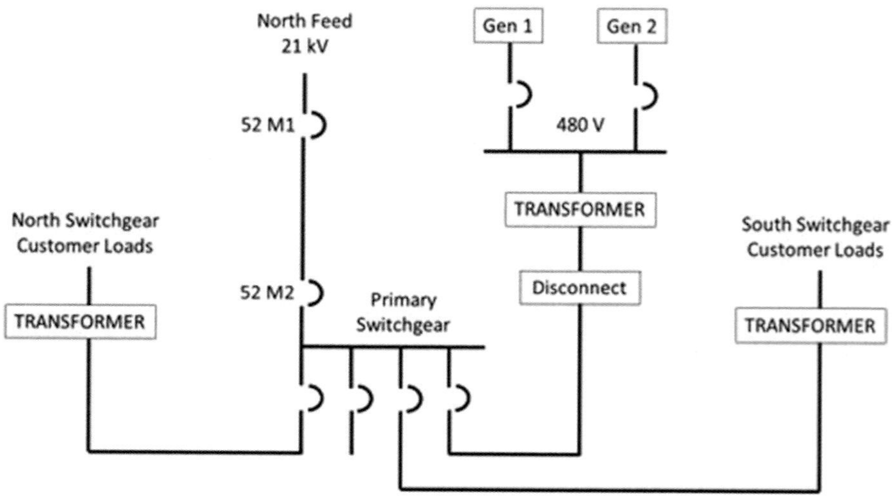

Fig. 7–14. New electrical distribution system

Yes, this means the electricity from the cogeneration power plant would have to go through a step-up transformer and then through a step-down transformer before reaching the various loads of the customer. The inherent inefficiency of this process was an acceptable loss to the overall system as a compromise to centralize the previously split infrastructure. The economics of reusing as much of the previous equipment as possible, as well as minimizing the required amount of downtime for the customer's facility, together with the possibility to increase the run time of the gensets resulted in a net gain for the ESCO. Also, combining the two original feeds from the local electric utility into one and bringing the central switchgear to 21 kV allowed the facility to shift to primary service from the utility, which had a lower rate structure than the previous secondary service.

The engine-generator sets were from a different OEM than the original equipment, and the generators had 100 kW more capacity (1.2 MW each). At least one of these generators could be run all through the winter off-season, and both units were expected to be running all summer. Each engine had its own HRSG and exhaust stack, with both a direct path and a post-HRSG path available, making emissions testing easier. Each engine also had a jacket water heat recovery system, increasing the overall efficiency of the power plant. Hot engine coolant went directly to a heat exchanger. After the heat exchanger, there was a temperature-controlled bypass valve to direct coolant flow to the dump radiator as needed.

Two rooftop evaporative coolers were used to maintain a comfortable environment for the gensets, as they were now enclosed in building—albeit one with spaces in between the walls to allow for plenty of ventilation. One (potential) problem with this layout was that the evaporative coolers were installed directly above the engines, with their discharge going down through the roof, right on top of the where the engine and generator meet. A problem associated with the long-term operation of evaporative coolers is that the float valve that controls the amount of water in the reservoir can fail (earlier in areas with especially hard water). When this happens, the water in the reservoir can overflow the tank—which, in this case, would then drain down right on top of the generator.

The heat recovery from the engine's jacket water was straightforward. The hot coolant exiting the engine was routed to a plate-and-frame heat exchanger and then to a three-way valve. This temperature-modulated valve then directed as much of the coolant to the dump radiator as needed, to protect the engine. The fan speed of the dump radiator was also temperature controlled, allowing for flexibility and redundancy in the engine protection from overtemperature conditions. This is illustrated in figure 7–15.

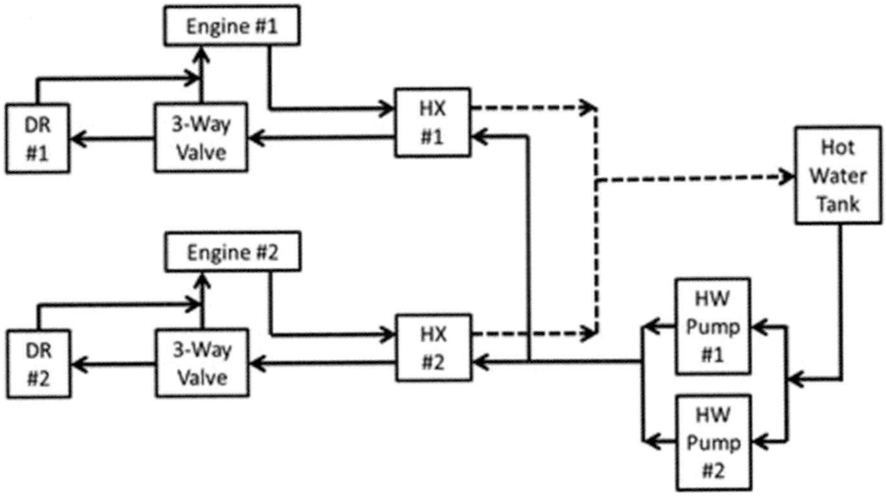

Fig. 7–15. Thermal harvesting of engine jacket water

A large tank of hot water was employed as a thermal storage system, and the food-processor customer could then use this hot water as desired in the facility. Water from the tank was pumped via one of two redundant pumps to the engine jacket water heat exchangers in parallel, with the now-warmer water (dashed arrows in fig. 7–15) returning to the tank. The engine's exhaust heat was harvested by the heat recovery steam generators only. There was still some viable heat in the engine's exhaust post-HRSG, but not much. One way to capture this heat is to use additional heat exchangers in the exhaust path, using chilled water as the heat sink. Even though the recovery of additional heat from the exhaust may be minimal, once the exhaust is cooled down enough the water vapor will condense out of the exhaust. For every molecule of methane that is consumed by combustion within the engine, two molecules of water are produced, and this water can be harvested, too.

Another way to get some use out of the low-grade post-HRSG exhaust heat is to use an *organic Rankine cycle* (ORC) system. This is the same basic system: taking a fluid, boiling it via waste heat recovery, using the vapor to drive a small turbine generator, condensing that vapor, and repeating the process. It's considered organic because it uses something other than water as the base fluid, usually a carbon-based molecule. A wide range of substances, usually thought of as being refrigerants, can be used for this purpose. There is an entire subindustry of ORC power systems, as an Internet search will reveal.

Common Failures

By chance alone, it is far easier to get an answer wrong on a test than to get it right. Some academic programs go one step further, eliminating even the possibility of a student making a lucky guess, by eliminating multiple-choice questions. All of the submarines in the U.S. Navy are nuclear powered, and the sailors that operate those nuclear power plants go through two years of training before being sent to the fleet. The only multiple-choice test in the entire program is the one taken to be able to join the military in the first place (the Armed Services Vocational Aptitude Battery [ASVAB]). In the case of those personnel operating nuclear power plants in a vessel submerged hundreds of feet below the surface of the ocean, I am certainly grateful that the academic path to get to that point is exacting. When shopping for an ESCO to bring on-site power to your facility, don't hesitate to question the academic history of the engineers involved in the process.

If you are in need of a new car and visit a dealership, do you buy the first car the salesperson shows you? Maybe so, but the more prudent method is to look at many different makes and models, including going on a test drive, before deciding which new vehicle best suits your needs (and budget). When an industrial facility's chief executive officer decides that their site needs its own power plant, the initial research quickly hits upon the major equipment manufacturers of engine-generator sets. The product sales teams for companies like Caterpillar, Cummins, and Jenbacher, among others, will happily try to convince the chief executive officer that their particular equipment is the best fit. All of the industrial facility's energy management problems will be solved, all at a great, low price! Successful sales agents are good at what they do, but that does not necessarily mean that they have the best combination of equipment for the customer's true circumstances (including any plans for growth).

The amount of buyer's remorse that I've encountered as I've inspected one failed cogeneration power plant after another has been more plentiful than I had ever expected. As equipment sales representatives shoehorn their particular product line into service, it rarely works out as well as the customer was led to believe. The cliché of the proverbial used car salesman has some truth to it. When shopping for a new car, most people have a pretty good idea of what their specific needs are. However, unless the chief executive officer is also a power systems engineer, he or she probably doesn't really know what the industrial facility's energy needs are. Just because you buy a skillet, it doesn't make you a chef. Similarly, just because a company decides to purchase its own power plant, that doesn't mean they will know what to do with it. For example, not knowing that extended operation at low engine load levels will bring about early failures to the equipment has lead to the demise of many cogeneration power plants.

Reflecting on all of the ways I've seen people screw up the design, engineering, construction, and/or the O&M of on-site cogeneration power plants (despite their best intentions otherwise), I've categorized them into the 14 groups described in the following sections.

1. Failure to evaluate all possible solutions to the situation

Failing to evaluate all of the possible solutions to the particular set of circumstances for any given customer is the single most common cause of the failure of an on-site cogeneration power plant. A detailed performance analysis—using realistic estimates of how the equipment will actually perform under real-world conditions and assessing the resultant impact on the financial performance of the power plant—is a minimum due-diligence effort. Once all the analyses are done, including sensitivity analysis, and all of the various design options are compared on an apples-to-apples basis, then a reasonably sound decision can be made. Later, if things do not work out as planned, then at least an effort was made at the start of the project. Without any review prior to breaking ground, there would not be much recourse for seeking consequential damages via litigation with the developer of the on-site power plant after a few years have gone by. From 1999–2003, especially in California, several small businesses were started with the idea of being an ESCO and bringing cogeneration power plants into existence for industrial sites and food processors eager to save money on their energy expenses. All but one of those initial developers are now out of business. Their designs were flawed, their engineering and construction practices were poor, and the power plants that were built have all failed. The ESCOs that followed have met with people in industries best served by having an on-site cogeneration power plant but who were poisoned to the idea, as industry associations had shared the lessons learned about the failures of such systems. Consequently, the market has been tough for more-recent ESCOs.

2. Failure to correctly identify the initial functions and requirements

This is similar to the first line of failure, but more specific. It could just as easily be "Failure to use a systems engineering process." The sales team from any of the big equipment manufacturers will happily confirm that their line of cogeneration power plant products will meet the needs of any prospective customer. Caveat emptor!

How is the customer supposed to know if the assertions of a sales representative are true? One common way is for the customer to hire its own *owner's engineer*, who whose responsibility is to understand everything that is going on with both the customer's facility and any concept for a new on-site power plant. Under the assumption that the customer is not a power systems engineer, the only way for them to catch any misleading promises by the equipment sales team is to have their own representative who speaks the same language. It is not uncommon for facilities like college campuses to hire an owner's engineer to assess their current situation in the first place, before making any requests for proposals from power plant developers.

One problem with this approach, however, is that the owner's engineer may put too much emphasis on the design configuration he or she comes up with. That is, if one of the power plant developers were to propose a hybrid solution, or some other variant of a power plant design that is not entirely similar to the one the owner's engineer suggested, the customer's team reviewing the proposal might discount it. I have seen project review committees succumb to tunnel vision, wanting to entertain only those proposals that match what their own engineer suggested, to the exclusion of all else. In their defense, these are typically not power plant people, so they are acting on the advice they have paid to receive and don't truly understand the nuances that some alternative design might take, even though it is just as valid a solution as the one suggested by their engineer.

In any case, regardless of how the performance modeling and financial analysis turn out, if the on-site cogeneration power plant was not designed and built to supply the demands of the customer, it will end up being for naught. If you're looking to buy a new car that you can take off-road, four-wheeling through the woods, then you will be disappointed with a small, electric-gas hybrid with a 100-mile range, limited to driving on pavement. Sure, the economics might be great, but it doesn't meet the needs you established from the beginning. So it is with power plants as well.

Using the systems engineering approach (described in detail in chap. 2) will mitigate the problems associated with *design creep*. That is, as the power plant developer goes through design and concept refinement, sometimes the target can drift. If the developer doesn't continuously refer to the agreed set of performance goals, it is easy to chase a design that, if built, will not deliver what was promised at the start of the project. This applies to the electrical distribution as well as the thermal outputs of the power plant. The case studies in this chapter highlight that a poor understanding of how the customer will actually use the electricity generated on-site can lead to failure and customer dissatisfaction. An Internet image search for "tree swing project management" gives a cartoon aptly illustrating the situation.

3. Failure to address dust and other site-related hazards

Even if a new on-site cogeneration power plant is beautifully designed, well-engineered and constructed, and operated by a crack team, it can still fail if the site-related hazards are not accounted for. The hazard that is the most underestimated is dust. Whether it is sawdust from a wood-based industrial facility or the dust generated by the various grains processed at a feed mill, it invariably leads to the early demise of any power plant subject to that dust.

It's not hard to imagine why this is so, especially if you've ever been in the middle of a dust storm. Machinery generally operates best in a clean environment. Blowing dust can become electrostatic and thus can cling to surfaces. Most people can run their finger along the front of whatever television screen they usually watch and leave a trail in the dust gathered there. The machinery in a power plant will also collect dust, and in the case of dust that is also conductive, bad things can happen to the inside of a generator where the electromagnetic fields can be very strong. Engine parts and the exposed surfaces of heat exchangers (e.g., tube fins of a dump radiator) can become coated in dust a lot faster than most people expect. Worse, with even a light rain or just very high humidity, sometimes that dust collection can set like concrete or adobe.

4. Failure to size the generation units correctly or operate them in the correct mode

Having the right tool for the job is only part of what's necessary; one must also use the tool in the right way(s) to accomplish the given task. So it is with the equipment used in a power plant. This is where having knowledge and experience comes into play. Modern engine-generator sets (as well as combustion turbines and the other technologies) all have very complex control systems. So, too, do the other equipment, like the protection relays and switchgear control systems, yet all of these systems need to work together in harmony.

If you've ever tried to cut lumber with an underpowered chainsaw or tow a trailer up a steep hill using a small car, you know the frustrations of trying to do something with the wrong-sized equipment. In the case of an on-site cogeneration power plant, having undersized equipment is not entirely bad. It's far better for the engine-generator set(s) to be running at full load than at 50%. The more common problem I've seen is an oversizing of the generating equipment. If the need is 360 kW, let's put in 560 kW worth; if the need is 3.6 MW, let's put in 4.4 MW worth; and so on. Often, this is done in anticipation of future growth of the electrical demands from the customer's planned expansion(s) of their facility. Unfortunately, these planned expansions are envisioned to occur many months or even years in the future and are often postponed in a perpetual exercise of shifting expenses to the right on the calendar. As a result, the combustion-driven equipment is operated at less than full capacity for extended periods of time, yielding a host of problems (as discussed earlier).

A modular approach is the system designer's friend. Using multiple smaller generation units will allow for the flexibility needed to operate the power plant at high levels of efficiency in the short term. Also, regardless whether the units to provide the anticipated extra capacity are installed at the start or purchased and installed later, the infrastructure can be built into the initial power plant to accommodate them. This is analogous to a young married couple buying a four-bedroom house. They won't be using all of the bedrooms initially but are planning on having children at some point in the future, so the extra bedroom capacity can remain dormant until needed.

Furthermore, once the right number of generators is determined, they must be used correctly. How many people have used a flat-head screwdriver as a chisel or a wrench as a hammer? The choice of which control scheme to use will be different for each on-site power plant, but care should be taken to use the best one. Does it make more sense to operate the generating equipment as a peaker plant or in a load-following mode? How should the power generation and distribution system respond to a loss of utility power? How should it respond once the utility power is restored? Some things can be determined, or at least be informed, by extensive modeling of the various system used. Other things will need to be well thought out via what-if analysis. Again, this is where the experience of the chosen ESCO's staff really shows its worth.

5. Failure to provide adequate engine operating environment

This is another category where hindsight is truly 20/20. Consider the idea of wanting to build a house. A modern log cabin might work well for a homesite in the Colorado Rockies, but that same design wouldn't fare so well in the swamps of the Louisiana bayous. So it is with power generation equipment. Feed mills are almost always a good fit for on-site cogeneration, given their need for both electricity and heat—but they are also very dusty places. Even when power systems engineers design for these environments, the as-built facility proves to be inadequate. Is this because of underestimating the extent and severity of the operating environmental challenge(s) or choosing not to spend the money on protections that are deemed too extravagant? Or is it both?

Whether it is using containerized units without sufficient ventilation, or open-air installations with inadequate weather protection, or designs based on seasonal averages rather than on the extreme conditions that have been observed at the site, the end results are similar. The given power plant will have O&M issues due to the environmental conditions at the site, and these will have negative impacts. Higher than anticipated engine derating, accelerated wear/embrittlement of hoses and wiring, rainwater getting into equipment not rated for exposure, clogged air filters, hard water deposits clogging drain lines, electronic control interfaces overheating owing to inadequate air conditioning in the control room—there are a host of environment-induced problems that must be anticipated.

6. Failure to model all parameters that reduce engine operations

When examining the equipment specifications provided by the OEM, many ESCOs I've dealt with have limited their studies to the statistics of full-capacity operations only. Similarly, in discussions on equipment selection with the sales representatives from the various OEMs, the focus is always on full nameplate performance of the given piece of hardware. Basing a performance model on these data alone is an exercise in wishful thinking. Unless the power generation equipment will be used only for baseload support, where it always be operating at full-load conditions, it will never perform as advertised. The information is available on the spec sheet for the equipment; nevertheless, this information is often ignored. The OEMs typically list values for all of the pertinent data at 100%, 75%, and 50% load levels, in either a table or a figure. Most of the time, the OEMs will also show graphs of the changes in the performance of the equipment as altitude, temperature, and other parameters change (e.g., fuel methane amounts).

A rule of thumb when choosing which data to base your model of engine performance on is to select the worst-case information over the optimum nameplate data. An even better approach is to include all of the OEM's equipment performance information in your model, using equations or look-up tables. In this way, for each 15-minute interval of time in the model, a much more accurate estimate of engine loading, heat rate, and other parameters of interest can be made.

For a complete understanding of the given situation, perform sensitivity analysis on the key areas, such as heat rate (fuel consumption). As the cost of the fuel changes by ±30%, how does that affect the financial performance of the power plant? When using more of a real-world approximation in the model, the results will assuredly be less attractive than when using full nameplate data alone—which is why equipment sales representatives will never do it that way.

7. Failure to be realistic about operations and maintenance

Equipment fails. Eventually, any piece of mechanical equipment will fail owing to wear on the parts. Electrical circuitry can be very robust, yet a lightning strike in the wrong place will fry things. I've seen the results after a red-tailed hawk spread its wings atop a power pole, only to become a phase-to-phase short that resulted in a transient-induced power plant shutdown. Rodents enjoy chewing on the insulation on power cables. In other words, there will be equipment downtime that is unexpected in the life of a power plant.

Using a performance model to estimate the number of run hours (and under what load conditions) for power generation equipment is pretty easy. There are preventative maintenance plans that use equipment run hours as a metric, rather than calendar days, and the time the equipment will be off-line can be estimated. One problem with this approach is the *downstream effect,* where the failure of one piece of equipment (for whatever reason) takes other equipment off-line, too. For example, if the only motor driving the fan of the dump radiator for the oil cooler of a reciprocating engine is unavailable, then the engine will be unable to run even though there is nothing wrong with the engine itself. The downstream effect is difficult to predict with precision. A conservative rule of thumb is that at any time there is a 10% chance that an engine (or combustion turbine) will be unavailable (for any reason).

As a hedge against situations where the failure of a $500 part could cause the whole power plant to be shut down, plan for such events. Being able to anticipate the components that are single-point safe is not difficult. Once these have been listed, stock replacements of the various parts that, on failure, will take that component off-line. Items such as belts, hoses, wiring harnesses, fuses, filters, spark plugs, relays, and igniters may seem trivial when designing a multimillion-dollar cogeneration power plant, but ignore them at your peril.

There are several different categories of equipment maintenance:

- *Anticipatory maintenance* comprises those repairs made to equipment because of the potential for sudden failure of parts or systems that are not routinely checked. If there are many failures of the same equipment, the manufacturer often calls for the faulty component(s) to be replaced early (e.g., automobile recalls).
- *Periodic maintenance* is just what the name implies: after a certain period of time (or run hours), the given maintenance item is to be performed. A common example would be changing the oil in your car.
- *Preventative maintenance* entails examining the various parts of a piece of equipment that can fail owing to wear—checking the extent of any problem(s) found and making repairs or replacing components as necessary.
- One variation (among many) of preventive maintenance is *predictive maintenance,* where a preventative maintenance item is triggered

when a certain threshold is reached. Temperatures, pressures, bearing wear, changes in pH, and limits on tank levels are common triggers for such work. In any case, when checking the condition of equipment, it is reasonable to expect that sometimes further work will be required, whether it's further investigation/triage or repair/replacement of the component(s). This is "predictable" only in a general sense—hence, only a 90% probability of uptime is used in the performance model (see fig. 5–7).

- Once a component has failed, *corrective maintenance* is performed.
- If the repairs are critical or otherwise urgent, *emergent maintenance* is required.

Some argue that maintenance and repairs are two separate tasks—just as eating right and exercising (preventative health measures) and having surgery (a repair to the body) would be considered distinct. To me, the important part is not in semantics but maximizing the equipment's uptime. Although engine overhauls are usually the only extensive periods of downtime for engines and turbines, it is prudent to allow additional downtime for routine preventative maintenance.

8. Failure to consistently apply all rules and standards to each alternative

This is a problem with human nature, not with any particular piece of equipment or system design. As part of the systems engineering process, whether strictly followed or merely approximated, all possible solutions to the given power plant problem are explored. For example, there will be cases where the use of RICE gensets with energy storage may be comparable to using a combustion turbine–based configuration with thermal energy storage. Exceptions to the rule should not only be expected but hoped for on the part of the developer of the on-site power plant. As with shopping for a new car or mattress, the more selection you have to choose from, the closer you will get to meeting your needs.

As humans, we all have preferences, and these come into play when selecting the best combination of equipment to use in a power plant. Whether these preferences come from a long history of experience working with particular equipment, or perception of quality, or any other reasons, many power systems engineers are biased in favor of specific equipment and/or the configuration/use of that equipment. This is called *enthusiasm bias* and is played out both in the analysis of the different approaches to solving the power and energy problem(s) of the given prospective customer and again in the comparison of those solutions. Enthusiasm bias can be summed up as the failure to consistently apply all of the rules and standards set forth at the beginning of the project to each design under consideration.

Preference isn't limited to areas where qualitative analysis is used, but shows up during quantitative analysis, too. Personal preferences guide the selection of margins, what threshold(s) are chosen, and how costs are defined, among other details. For example, if the model using a combustion turbine was analyzed using natural gas at $4.00 per million BTU (MMBTU) but the model using RICEs was analyzed at $3.50/MMBTU, then the results when comparing the two solutions will be skewed. This often happens when there is more than one person performing the analyses: One person will use one set of constraints (like the cost of fuel) while another will use some other factor, or even two different units of measurement (e.g., MMBTU vs. cubic feet).

One problem with the landing of the Beagle 2 probe on Mars in 2003 resulted from such a failure to standardize. The mission was a joint one between the European Space Agency and entities in the United States. The U.S. engineers used standard units, while the Europeans used metric ones, and no one caught the different use of units of measurement, paying attention only to the quantities themselves. The probe was ultimately lost, although other problems contributed to this, in addition to systems engineering failures.

9. Failure to include life-cycle costs in financial analysis

One frustration power systems engineers face is that, when presented with a design for a new power plant, accounting personnel look only at the short-term financial impacts. That is, their focus is on the net dollars per kilowatt of installed electrical generation capacity and other, similar metrics of the proposed power plant. If a proposed facility would exceed a predefined limit, such as $800/kW after everything was installed, then it becomes a nonstarter, regardless of other concerns. The total life-cycle costs of the project and the net revenue it could provide the ESCO via a PPA are discounted (if not ignored) because the accounting personnel will balk at the initial expenses of building the power plant—even when the net revenue is projected to be large. I've seen it happen.

Another example of failing to include the full life-cycle costs of the power plant in the financial analysis is when the costs associated with the overhauls of the equipment are not taken into account. Rather, using the nameplate data of the equipment and average annual production estimates, the net cost-benefit analysis of the power plant's lifetime is skewed from reality. The two biggest drivers of the financial viability of a proposed power plant project are the heat rate of the prime movers (fuel usage) and the costs associated with the necessary periodic overhauls of the equipment (including the associated downtime).

This is similar to deciding not to buy a particular car because its price is beyond your threshold of acceptability, even though the vehicle was projected to get 200 miles per gallon of gasoline consumed. Over the life span of the car, the savings in fuel costs would more than offset the higher than desired initial price, yielding a net gain in your personal finances. The purchase is never made, though, owing to the unacceptable short-term financial impacts. Such protectionist thinking has left many perfectly viable power plants on the drawing board, never to see the light of day.

10. Failure to perform or understand the results of sensitivity analysis

Sensitivity analysis is a way of checking the effects, on the system being modeled, of changing one variable while holding all the others constant. The range of variability is usually at least ±20%, but sometimes it's reasonable to have a wider range. Plotting the results on a graph highlights which other parameters are most affected by the change in a given parameter (fig. 7–16).

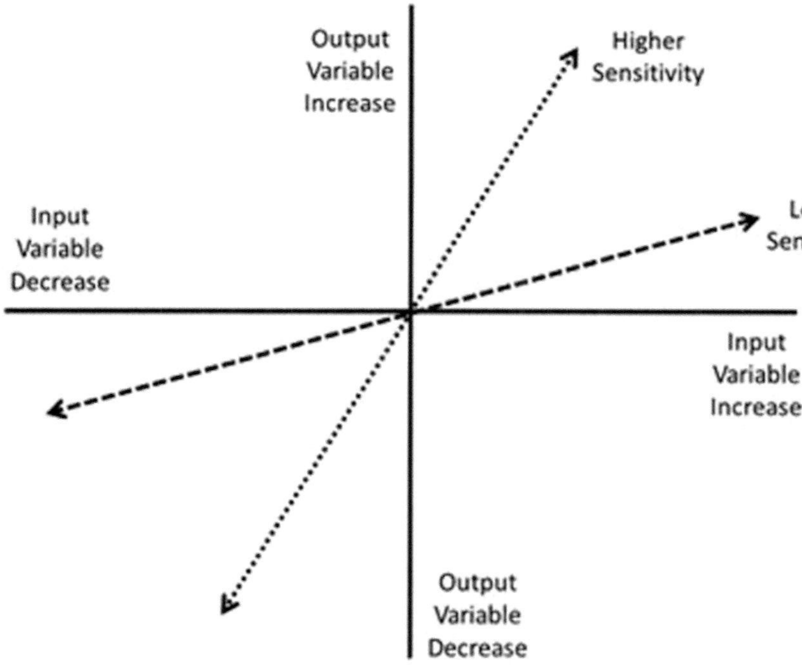

Fig. 7–16. Sensitivity analysis

Consider the effects on everything associated with the financial performance of a the power plant being modeled if the price of the fuel were to vary by ±30% (i.e., the range of input variable increase and decrease). When the change in the costs associated with the maintenance of the prime mover as the price of the fuel changes is plotted, there should be a very low-sloped curve (low sensitivity). Plotting the resulting change in the monthly costs associated with the operations of the power plant should be more severe (as indicated by the higher slope), since the costs associated with the fuel consumption are a component of the total operating costs of the facility.

Although some relationships that sensitivity analysis elucidates are common sense, others are less so—and even if the resulting sensitivity was predictable, the magnitude of that sensitivity may be surprising. For example, what would be the effect(s) on the performance of the power plant if there were 20% more sunny days over the course of a year? How would the increased output from the given PV component affect the number of cycles of the energy storage system (battery charges)? How much less fuel would the prime mover(s) need to consume? Would the decrease in run time affect the number of oil changes required?

By performing multiple sensitivity analyses on every what-if situation you can think of, you will discover the weaknesses (and the strengths) in your design.

11. Failure to use realistic probability distributions in Monte Carlo analysis

While a sensitivity analysis changes one variable at a time, holding all other constraints constant, in a Monte Carlo analysis, all of the variables are allowed to assume any value within their set ranges. The key is to assign realistic ranges for the probability distribution curves for each of the parameters of interest. Just as important, though, is to have a computer program or spreadsheet application that will run such an analysis. The "Solver" function in a spreadsheet program can be used to adjust selected variables to produce a desired result—for example, minimizing a given cell (e.g., net imported power from the utility). This is helpful, but goes only so far. By the development of a range of possible values for each of the variables at hand—even the amount of time it will take to construct the power plant—a full analysis can be made of the given project.

In the simulation, a value is chosen at random from the defined range of values for each of the parameters under analysis, and the model is run, with the results going into a new data set. This process is repeated thousands or tens of thousands of times, to yield realistic, increasingly refined results—hence, the need for a computer. The data set of results will show the likely probability of the many outcomes under study, from the worst-case to the best-case scenarios. Taking the results of the analysis for each parameter and plotting their distribution on the basis of the number of times the results were achieved yields a probability curve. Nevertheless, as with any other forecasting tool, the results will be only as good as the estimates used to define the ranges of the variables used in the calculations.

Another useful what-if tool is Bayes' theorem, which is a way of working with probabilities of probabilities. For example, what's the probability that the generator end of genset 1 is out of commission at the same time the engine end of genset 2 is down? There can be a conspiracy of multiple, unrelated equipment failures that, if they happen at overlapping times, can seriously reduce the capacity of the power plant. But how probable is that? Having redundant systems in place to ensure the absence of any single-point safe condition will minimize this probability. The problem is that when one of the units in a redundant system is off-line, for maintenance or repair, the redundancy is lost for that particular system, and a single-point safe condition can exist.

Using mathematical models of probability analysis is a good way of finding weaknesses in the design of a power plant. However, these models are only as good as the inputs given. Garbage in will equal garbage out.

12. Failure to document everything, including why a system was selected, for later reevaluation

One of the simplest things to do while plodding down the power plant development path is to take notes. As the systems engineering process progresses and the various design options and/or equipment selections are rejected, it is very handy to have notes detailing the reasons why such decisions were made. This will prevent the otherwise inevitable question at the latter stages of development: "Why did we choose that?" This is invariably followed by digging through old e-mails—and wasted time—to verify that yes, it was a good decision to reject it.

Considerable second-guessing and rehashing of old discussions happens when the stakeholders get together for a discussion of the final selection of the details so the project can move toward procurement and construction. Personal biases get raised—for example, someone thought a different type of equipment and/or arrangement of equipment was going to be used. This happens when the project development group does its job without continuously giving feedback information to all of the other stakeholders in the project. Everyone needs to know why the equipment were selected instead of other available equipment (and configurations), to avoid unnecessary delays late in the process as a result of having to verify why matters were done as they were.

13. Failure to identify the customer's true needs

Of all the ways an on-site cogeneration power plant can be screwed up, this is by far the most common. Understanding the problem before trying to find solutions for the problem is a critical part of the process. Will the heat exchangers used need to be NSF rated? What are the seasonal energy demands of the prospective customer? How realistic are the plans for future expansion, and will that expansion change the electrical-thermal usage coincidence? Does the prospective customer need more steam, or will hot water work instead? Cutting corners and speeding things along during the initial stages of the systems engineering process will not prove beneficial in the end.

Those ESCOs that already have several cogeneration power plants in their resume, as well as the salespeople from the various OEMs, will have to fight the temptation to use the one-size-fits-most approach to the prospective customer's set of energy management problems. Eliminating the months of analysis and systems engineering (thereby saving a ton of both time and money) by using one of their preexisting solutions is something most such developers cannot resist doing. It's hard to blame them: They already know what it will cost them to build, how long it will take, and what kind of performance they can expect from the equipment—all without spending a dime on engineering research.

Nevertheless, the degree of uniqueness associated with each new application of distributed generation should not be underestimated. Every potential customer will have a different set of circumstances, needs, goals, and plans for future expansion, and these must be evaluated individually if any precision is to be had in solving the customer's energy management problem(s). Still, it's possible that preexisting solutions may fit a new prospective customer's needs, and a little tailoring can address any differences to better match the solution to the problem. This is like buying a new suit; most people start with the selection of existing trousers and jackets, and upon finding one close to the right size, a bit of tailoring can resolve any issues in fit. Caution is warranted when starting with such preexisting solution, though, because more than one "fit" may need to be made when dealing with an on-site power plant. For instance, solving both the electrical and thermal energy needs of a prospective customer with a readily available package of equipment would be a coincidence that should warrant skepticism on the part of the customer.

14. Failure to perform failure analysis

Why did the equipment fail? Why is the heat recovery less than design estimates? Why are flow rates less than design values? One of the more frustrating causes of failure in on-site power plants is that, as the saying goes, Those who fail to learn from history are bound to repeat it. That is, ESCOs and other power plant developers have been known to reuse system designs and/or equipment that have not worked well in the past. Even though most people are usually averse to repeating what they know to be a mistake from personal experience, the same is not always true of OEM sales representatives or power plant developers—after all, this new project is a new set of circumstances, so perhaps the equipment will work in this case. Unfortunately, it rarely does. When a power plant performs suboptimally, or equipment fails prematurely, or the system design fails to perform as expected, there is an open invitation to learn: Why did the equipment fail? Why are flow rates less than design values? Why is fuel consumption so much higher than forecast?

Performing failure analysis on equipment and/or systems that do not perform as well as expected is crucial. The ESCO should use their new understanding of the operation of existing equipment to improve future designs; this represents a growth opportunity not to be missed. In preliminary meetings with prospective ESCOs, I'd much rather deal with a power plant developer who admits to less-than-perfect success in past projects, especially if they go on to describe how they have become a better ESCO by learning from their mistakes, than one who claims to have a perfect record.

8

Concluding Points

Benefits of On-Site Cogeneration

There are many benefits to any facility's having its own on-site cogeneration power plant. The basic driver of the decision to employ this technology is financial—buying power wholesale, rather than at retail prices. Although the economics are certainly important, there are a wide range of other advantages to consider, the relative value of which will vary depending on your unique circumstances.

Reliability

When done correctly, installing an on-site power plant can provide very high levels of reliability for the supply of electrical and thermal energies. By use of a combination of a modular approach with backup systems in place, such as energy storage systems, sufficient redundancy can be achieved to securely provide continuity of power to the customer. Having multiple smaller generation assets will spread the risk posed by individual equipment failures, but at a cost. The economy of scale of engine-generator sets is such that, on a dollars-per-kilowatt basis, larger units are less expensive than smaller ones. There is also a point of diminishing returns where the costs associated with increasing the reliability exceed the potential benefit. Having $N + 3$ generation assets is handy, but how much more expensive is that compared to having only $N + 2$? Is the extra cost of another genset worth the small improvement in overall system reliability?

There will always be limits and vulnerabilities to consider. Electrical transmission lines from the nearest utility are vulnerable to forest fires, tornadoes, and other natural disasters. Pipelines carrying natural gas can fail because of accidents due to human error or large-scale events such as earthquakes. Being off grid(by using locally available hydroelectric power, solar power, wind, biomass, etc.) can increase the overall reliability of the on-site power plant. Together with battery systems and/or thermal energy storage systems, any on-site cogeneration power plant can be as reliable as desired.

Economics

The cost of electricity is very high in some parts of this country, and it isn't likely to decrease soon. As population centers grow, so too does the demand for electricity. Unfortunately, most electric utility companies do not have large reserves of capacity, nor do they have excess capital available to construct new power plants and transmission lines. Instead, small, modular *peaker* power plants are being constructed to boost the power generation capacity during periods of high demand. Concurrently, utilities are offering incentives to businesses and homeowners to curtail electricity use during peak demand periods, including giving control of their appliances (e.g., air conditioning) to the power company. This supply versus demand situation is driving many homeowners and business owners to embrace the idea of the installation of small solar power plants on the roofs of their homes and businesses. Those businesses consuming at least 350 kW of electricity (and have a thermal load) should also be examining the lower rates from having their own on-site cogeneration power plant.

Generating your own electricity, for less money than buying the equivalent amount of power when purchased from the local electric utility, is pretty straightforward. Getting the thermal energy from the cogeneration power plant at a rate cheaper than making it themselves is another plus (think boilers and chillers). Less obvious economic benefits include having lower maintenance costs. This is because the electricity generated on-site is cleaner than electricity from the grid. That is, there is less distortion of the electrical sine wave. Voltage, frequency, and stability are all improved, which leads to an increase in the mean time between failures of equipment such as pumps and motors. While difficult to forecast, these savings will be realized after several years of operation of the on-site power plant when maintenance costs will be less than when the facility was using electricity supplied by the utility. Similarly, the economic benefit of having increased continuity of power is difficult to forecast. Every time there is an interruption in power from the utility, what is the cost to the customer? Lost product, lost production time, idle personnel, increased maintenance costs, and the consequences of a loss of refrigeration are all examples of financial impacts.

In businesses where the margins are thin, being able to predict with certainty what the budget impacts will be from the use of energy is a clear advantage. Generating electricity and thermal energy on-site with a long-term contract for the purchase of the fuel can lock in the price of power for years at a time. The rates charged by the local electric utility will vary and can be increased at any time. Having direct control over the on-site energy production and usage allows any facility to maximize their operational and economic efficiencies.

Green

One popular benefit of getting electricity and thermal energy from an on-site cogeneration power plant is that it is very environmentally friendly. Utility-scale power plants can operate at efficiencies as low as 30%, and then there are line losses as the electricity makes its way across the transmission and distribution systems. Cogeneration power plants can easily operate at 80% efficiency, and many of them operate higher than that, depending on the heat recovery systems. These higher operating efficiencies translate into a net fuel savings compared to using utility power. Burning less fuel represents a net reduction in the greenhouse gas emissions. Having a tight condensate return system also saves on water usage and increases operational efficiency. The water that is boiled in the HRSGs, after being used as steam, can be condensed and returned to the HRSG. If using a condensate return tank, it can feed the deaerating feedwater tank instead of using fresh makeup water. Already at a high temperature, less energy is then needed to bring the water back to a boil, and further deaeration can occur while using less steam in the deaerating tank. Having a CO_2 scrubber on the final emissions stack, such as the one made by Sidel Systems, together with a water vapor condenser system (also made by Sidel Systems), makes the entire process very environmentally friendly.

Easy, but not simple

Even though there are numerous potential problems with having an on-site cogeneration power plant, it's not all negative. Although there are dozens of factors to consider—each with negative consequences as a possible end result, some quite severe—an on-site cogeneration power plant can be an elegant solution to a facility's energy needs. The process is exacting, but it's still relatively easy to do. There's almost always more than one way to solve a given set of energy demands.

The problem is that most power plant systems are assembled à la carte. The technology isn't new, but the pieces of the puzzle have gotten more sophisticated. Sure, you could build a car in your garage from individual parts, but it probably wouldn't be as good as those made by folks who've been making them for years. To mitigate the chance of missing something, make a checklist of items to consider and their relative importance. This should include items of concern such as the following:

- Make sure comparison studies among possible solutions are comparing apples to apples.
- Always compare the final set of possible solutions to the do-nothing option.
- Compare the cost of supplying electricity and thermal energy to the cost of the absence of these energies to the customer—in other words, put a price on reliability.

- Customer needs can change. Anticipate 2, 5, and 10 years of growth. This is where a modular design concept has the advantage. Design the systems and system interfaces with future expansion in mind. This also applies to the original footprint of the power plant.
- Ensure that the ESCO is willing to provide some level of training to the customer (on O&M issues, emergency procedures, system operations, etc.) even if the customer is not expected to ever operate the power plant.
- Float power purchase agreements on five-year contracts—significant changes to commodity prices can take place in that time period. This also provides an exit ramp for the customer if the power plant has failed to meet expectations.
- Establish a spare-parts inventory and an inventory-control system. Anticipate long lead-time items using the preventative maintenance plan.
- Measure temperature, pressure, and flow by using supervisory control and data acquisition (SCADA), to anticipate changes in the flow through all heat exchangers. The heat transfer surfaces should be kept free from fouling, using filters and chemical corrosion controls.
- Include centralized valve operating manifolds and operational stations of fluid systems where possible. Avoid installing valves in hard-to-reach places. The U.S. Department of Energy and other organizations have guidelines on the design considerations for power generating facilities.
- Note precautions in the construction plans for operating the power plant in inclement weather, including having sufficient lighting.
- Consider using air-lock access systems and/or a positive-pressure system for containment buildings in high-dust environments.
- Take extraordinary precautions against cyber-threats for remote operations or human-machine interface access.
- If using natural gas-fired boilers in the facility, consider using a microturbine as the blower.
- Ensure that air-filter square area is based on fouled-air filter flow, not clean.
- Hybridize when possible, using a mix of the available technologies and sources of heat and power.
- Find ways to use organic materials currently in waste streams (biomass).
- Closely examine all of the supporting utilities (compressed air, chilled water, etc.) for possible support from the cogeneration power plant.
- Include a calibration program. Temperature, pressure, differential pressure, and other indications are not valid if not verified periodically.

Build a Successful Team

A beautifully designed cogeneration power plant won't be able to perform well if the engineering is poorly done. A well-engineered, beautifully designed power plant won't perform well if it's poorly constructed. And a well-engineered, beautifully designed, well-constructed power plant won't perform well if the O&M crews are unqualified, poorly trained, or simply inept.

Experience counts

Finding a reputable, experienced design and engineering firm and then construction contractors with a construction manager capable of shepherding all of the subcontractors is challenging. Getting a complete commissioning of the equipment, including the written reports of everything that was tested, is also challenging; most commissioning is minimal, not extensive. Nevertheless, in many cases of failures in a power plant, the fault or weakness could have been discovered early, during the commissioning process, had the equipment testing protocols been more thorough. In addition to the design, engineering, and construction companies, OEMs can assist with plant commissioning (which can fill gaps in the testing protocols). All of these companies should also be involved in the training of the O&M crew. Getting the O&M crew involved with the project early, even before construction begins, is extremely valuable to the health and welfare of the future power plant. There is no substitute for knowing the details of how the power plant systems are installed and interconnected, not just for routine O&M but also for troubleshooting faults when they arise.

Putting together a great operations and maintenance team

In general terms, the best O&M teams will encompass three groups of people: electricians, mechanics, and electronics/computer technicians. The electricians operate and maintain the electrical components of the power plant, including the following:

- Generators
- Voltage and current regulators
- Circuit breakers and switchgear
- Batteries and their support systems
- Wiring and cabling
- Motors
- Lighting

Similarly, the mechanics operate and maintain the mechanical equipment:

- Pumps and valves

- Hydraulic systems
- Lube oil systems
- Heat exchangers
- Air compressors
- Steam traps
- Piping
- Heating, ventilating, and air-conditioning systems
- Ammonia/urea and emissions-control systems

The electronics/computer technicians operate and maintain the "brains" of the power plant:

- The human-machine interfaces (HMIs), including engine-generator control systems
- The programmable logic controllers (PLCs)
- The direct digital control (DDC) system
- The IPS and GPS protection relays
- The computer-based control systems, such as SCADA and Tridium
- The computerized maintenance management system (CMMS)

One key to successful O&M is to ensure that these groups of people operate as a blended team, rather than as individual tribes. This is easier to do when there has been cross-training among the disciplines, so that members of each group can helping with (and learn about) the work done by the other groups. As learned in World War II, esprit de corps and morale are invaluable ingredients to maximize the effectiveness of personnel under stressful conditions. Although day-to-day O&M may not be especially demanding, troubleshooting, repairing, and restoring to service the critical infrastructure of a power plant to maximize uptime can be very stressful.

Leadership of the O&M team is also vital to the overall management of the O&M of every power plant—and, again, experience matters. Granted, there are some industries and commercial facilities that require managerial experience for their managerial positions, but not necessarily any direct experience with the given industry at hand. Power plants are not one of those entities. There is no substitute for the experience and expertise gained by working in a power plant for many years (and the corresponding exposure to a wide variety of the circumstances that come up). When screening applicants for O&M supervisory positions, I look for at least five years of O&M experience on that particular power plant, or at least eight years of experience in O&M overall. Even though this book is not about management or leadership, the importance of finding the right people to handle the management of the power plant and provide leadership of the O&M crew cannot be overstated.

Training and qualification programs

Just as important as finding the right crew to operate and maintain the power plant is what to do when you've found them. I have seen innumerable power plants where, once a person is hired (being deemed qualified for the position by previous experience), the only training they receive is on the job. The new personnel shadow current members of the O&M team until the supervisor becomes confident they are ready to work on their own, and that's it. Before long, these people become the mentors for the next new hires, and the process repeats.

As the senior personnel retire (or leave for other pursuits), their accumulated knowledge of the power plant goes with them. Over time, their actual understanding of how the power plant systems function is diluted (or lost entirely), and the O&M crew becomes ineffective—able to bumble through the daily routine, but unable to troubleshoot unusual circumstances. One or two people, who may be on different shifts, have some small amount of experience with the system having problems, but the personnel on duty don't have a clue. In the general power plant vernacular, this is known as *tribal knowledge*, and it represents an all-too-common state of affairs.

The way to avoid such reliance on tribal knowledge is to have a robust and formal training and qualification program for all of the power plant's personnel (including management). Establishing such programs is time consuming, but it has to be done only once; after the programs are established, they are self-sustaining. Most of the work should be done during the construction and commissioning phases, since that is when the details about how the systems are installed and interconnected are made known.

In addition to academics, the training program consists of the administrative and practical applications (i.e., actually performing some task). Typically, there will be a training manual (or other formal documentation) that describes the training and qualification programs in detail. This documentation is very important, especially if there will be consequences for those individuals who fail to make the minimum progress toward qualification, continuously fail the training examinations, or otherwise do not make good on the investment in their ability to do the job. If someone is to be assigned remedial training or even face employment termination after failing the training, then the human resources department will need written instructions that govern how everything is done. The document should give details including how and when the training is to be done, who is qualified to give the training, and what is expected to be learned by the trainee; thus, the details of any qualifying examinations, both written and oral, need to be specified as well.

Readings, classroom discussion and lectures, simulators, and hands-on experience should all be a part of the mix—with periodic examinations to measure retention (and accuracy of that retention). As the personnel become qualified on the various systems of the power plant, some may choose (or be appointed) to become subject-matter experts (SMEs), who can then give system-specific training to others in turn. What qualifies someone to be an SME on any given system should also be delineated in the training manual.

The qualification process that I recommend is a structure of formal study that happens concurrent with some on-the-job training. Once a newly hired power plant operator has completed the orientation process, they are given a *qualification card* and a copy of the *qualification standard*. On the front of the qualification card there are specific assignments, as shown in figure 8–1.

Fig. 8–1. Qualification card front

Training Area	Required Date	Date Completed	Authorization
Safety			
Systems			
Operations			
Emergencies			
Relations			
OSHA			

In the form given in table 8–1, the person's supervisor (or head of the training department) can assign the dates that the sections are required to be completed by, note when they were completed, and sign to acknowledge completion when the individual has been authorized. Similarly, also on the front of the qualification card will be the information about the required examinations, as shown in figure 8–2.

Fig. 8–2. Qualification card examination information

Examination Area	Required Date	Date Completed	Authorization
Safety			
Systems			
Operations			
Emergencies			
Relations			
OSHA			
Senior operator review			
Operations chief			

As can be inferred by the sequence of examinations in figure 8–2, the qualification card is broken down into the areas of operation of the power plant, with safety being at the top of the list. Safety encompasses electrical safety, equipment safety, chemical safety, and environmental safety. Under each of those headings there are several *checkouts* specified, with a place for a qualified person to sign and date the qualification card noting the completion of that specific checkout, as shown in figure 8–3.

ENVIRONMENTAL SAFETY

1. High-heat environments

Signature and date

Fig. 8–3. Example qualification card checkout completion

The requirements that must be met before getting an SME's signature will vary for each checkout and are specified separately in the qualification standard. These requirements generally include drawing a sketch of the given system (the level of detail determined by the qualification standard and the given SME), as well as verbally describing the various system components. An example of the environmental safety and electrical sections is given in figure 8–4.

Fig. 8–4. Example qualification standard

ENVIRONMENTAL SAFETY

1. *High-heat environments*. Describe all precautions for working in high-temperature environments. Describe the effects on the human body of light and heavy work in temperatures from 90°F to 120°F. Describe the physiological and environmental factors that induce heat-related injuries. Describe the methods of prevention, symptoms of, and treatment for cramps, exhaustion, heat stroke, and heat syncope.

ELECTRICAL

1. *AC power distribution*. Draw a one-line diagram of the system. Describe the purpose and operation of all components used in the system. Describe and give the set points for all alarms, warning lights, and indications, both locally and remotely. Describe local and remote operation of each component. Describe the safety features of each major circuit breaker and their set points. Describe the control power for each breaker. Describe proper safety measures and appropriate personal protective equipment required for safe operation of all system components. Describe the effects of weather on the operation of the system. Describe all maintenance requirements.

From the preceding examples, it should be clear that this is a daunting process, with over 100 systems to get checked out on and an intensive level of detail to internalize. Together with examinations to verify that this knowledge has been retained by the person being trained, the net effect is that it takes a newly hired power plant operator about six months to finish the qualification card (with all of the ongoing on-the-job training that's happening concurrently). At the end, though, after successfully passing a final oral board examination by the SMEs, the O&M supervisor and the plant's chief engineer, the person is immediately a qualified and productive member of the team. With routine training on a continuing basis, all O&M personnel can maintain their understanding of the power plant's systems, their operation and interconnections, and other facets. By devoting a short amount of time each week to doing some training/review, the O&M supervisor can mitigate the knowledge half-life that happens to everyone—because it is true that if you don't use it, you lose it.

Excellence in Operations and Maintenance

For a level of excellence to be obtained in O&M, a power plant must first have the right mix of personnel in terms of training, experience, and expertise. In my opinion, which I will grant is biased, the best source of training and experience for power plant operators (of all three desired types mentioned above) is the nuclear power program of the U.S. Navy. The enlisted personnel go through a two-year training program before they ever see an active aircraft carrier or submarine (all of which are nuclear powered). This consists of first learning a trade: mechanical, electrical, or electronics technician. Next, the U.S. Navy uses a four-year nuclear engineering degree curriculum (stripped of any fluffy nonengineering bits like art history, etc.) and force-feeds it to the students over a six-month time period with a cumulative examination at the end; this has been compared to trying to take a drink of water from a fire hose. Those who successfully pass the academic coursework then move on to a prototype nuclear power plant to get hands-on training (because Admiral Rickover, considered the "Father of the Nuclear Navy," did not believe in using simulators, which were fine for pilots but not for those who would be operating a nuclear power plant). This hands-on on-the-job training lasts another six months, with a final, cumulative written examination followed by an oral board examination. Successful graduates then go to the aircraft carrier and submarine fleets, where they begin the process all over again, qualifying to be able to operate and maintain their assigned ship. The initial training process takes about two years, and then the operators spend at least four years on their ship operating and maintaining the power plant's equipment.

Regardless whether these veterans leave the U.S. Navy after their six-year tour is complete or stay in the service through retirement, they eventually make the transition to civilian employment. Often, they will use headhunting companies (e.g., Orion International) that cater to military personnel making the transition to civilian employment to find their first post—active duty job, and this includes power plant operators. Many go on to staff the nation's nuclear power plants, whereas others go into the conventional power field.

In addition to the nuclear-trained personnel, other U.S. Navy personnel hold particular interest to those staffing a new cogeneration power plant. These include personnel trained and experienced in O&M of large stationary engines (engineman) or gas-fired combustion turbines (ground support equipment [GSE] or gas turbine electrical technicians). If I were tasked with staffing a new cogeneration power plant from scratch, I would immediately seek out the military headhunter companies like Orion International.

Communication

Even with the best and the brightest O&M personnel, some formal communication protocol is needed. This is often handled one of three ways:

- Posting an O&M schedule
- Keeping a detailed log
- Having (and maintaining) an accurate overall status board of the power plant

Posting a copy of the planned events for the power plant is pretty simple. A daily schedule, as well as one for the upcoming week, month, quarter, and year will keep everyone on the O&M staff aware of what's going on and what the overall operational plan is.

A detailed log is invaluable to maintain continuity of information from one O&M shift to the next, . These logs also serve as a legal record of the activities at the power plant (if ever needed), in addition to being a chronological history. Anyone should be able to review the entries in the log from the past 24–48 hours and know what is going on in the power plant (and why). The U.S. Department of Energy provides guidance on what should be logged (and other details about log keeping).

An overall status board provides a systemwide view of what is going on in the power plant. This can be a laminated drawing of the systems (or an overall piping and instrumentation diagram) that is updated using something as simple as a grease pencil. A more modern approach is to use a large-screen monitor to display the graphics from a building automation system (e.g., Tridium), including equipment status and alarms, among other details. This type of status board works well until the power goes out—and a complete loss of all AC power is an event that can happen, even at an on-site power plant.

Computerized maintenance management and preventive maintenance systems

Although they are related, CMMS and preventative maintenance systems (PMS) are not identical. CMMS is a computer-based, database-centric system to keep track of the equipment in the power plant. Each asset is recorded in the system, including the details of the make and model number, date when installed, detailed description and/or specifications. There are also fields for describing the various preventative maintenance items for each piece of equipment that are populated by information straight out of the technical manual.

The coordination of when to perform the various maintenance tasks is the tricky part, given that the database typically keeps track of the dates on the calendar but not the run hours of the equipment. Certainly, there are PMS items that should be done on a daily, weekly, or monthly basis, but most of the more serious maintenance items are driven by the actual use of the equipment (commonly determined by the number of run hours). Unless the CMMS is tied to the power plant's SCADA or building management system, these more important semiannual and annual maintenance items will not be managed by the computer.

Most of the time, the preventative maintenance schedule is dictated by the OEM in the equipment's technical manual, along with descriptions of the tools required to perform the work. Other factors will influence the scheduling of certain maintenance items besides the number of run hours, such as observations made by the O&M staff. The bearings of a motor will start to make noises and/or vibrations long before they fail, and during the routine checks of the equipment (by a regular staff), such unusual sounds and smells will alert the O&M staff to irregularities in the equipment that need attention.

The manufacturer's warranty is also a good clue with regard to timing of preventive maintenance. The OEM's reliability engineers will have determined at what point, in terms of time in service or run hours, the probability of failure is beyond their threshold of financial pain to cover. This can warrant increased attention to the care and feeding of the equipment, including increasing the periodicity of routine preventative maintenance items.

One constricting element when using a CMMS to organize the PMS of a power plant is the planning of outages (including coordinating these events with the customer and the local electrical utility). Taking an engine-generator set off-line to perform some kind of maintenance (such as an oil change) during normal business hours means an increase in the power supplied by the utility, thus an increased cost to the customer for that electricity and an overall reduction in the reliability of the power plant. This drives such routine work to be scheduled during off-peak hours (e.g., evenings and weekends). If the work is OEM or vendor supported, then the cost to the ESCO is increased accordingly for the technician's time outside of normal working hours. This also reflects back to the planning and the design of the power plant, in the avoidance of any single-point safe situation, where taking any one piece of equipment off-line for service will not limit the customer from full production capability. Most industrial facilities will have some kind of scheduled outage for maintenance purposes, usually annually around the Christmas holiday season. For the on-site cogeneration power plant, this is an opportunity to perform multiple PMS items at once and/or perform the more extensive maintenance items (e.g., an engine overhaul).

Another element of a good PMS program is maintaining an inventory of spare parts. Some engine parts that will fail are easy to predict (e.g., spark plugs) but the level of penetration can be a challenge—that is, do you include a spare of every part in the unit? Some OEMs will have a level of support with a warehouse of parts readily available to the ESCO, and some parts will not be on a shelf in any warehouse anywhere in the country.

If the OEM's main manufacturing plant is abroad, then the delay is getting certain parts can be a very long time (six weeks or more). Such less than readily available parts are usually the larger, more expensive ones that are also less likely to fail in the first place. Estimating the probability of failure can be difficult, especially for new equipment that does not have a long operating history for that particular OEM. Deciding in advance to purchase these expensive parts that are not likely to fail just to see them sitting on a shelf at the power plant are a manager's nightmare—and the company's financial officer will need to be convinced it's a good idea to spend money on parts that are not an immediate (or even foreseeable) need. The key is to balance the relatively low probability of failure against all of the costs associated with the equipment downtime while waiting for the part, including lost revenue for the ESCO and higher energy costs for the customer.

Quality assurance and quality control

A robust quality-assurance and quality-control system is a must-have for the successful implementation of any CMMS or PMS program. Just because a maintenance item is scheduled and the technicians are available, qualified, and resourced to perform the work does *not* mean the maintenance is done well. Were the air filters installed backward? Were the bolts tightened to the correct torque? Was the maintenance actually done, or just signed off as being complete?

In many industrial and power plant facilities, the maintenance staff will be working alone (or with no supervision). Unfortunately, not all maintenance workers are honorable people, and without a quality-assurance and quality-control program, there is no way for the management of the facility to know about idleness or other transgressions. Random, routine inspections by supervisory staff of the work performed (and documented) by the maintenance staff will assure that some minimum quality of workmanship is maintained. For example, most people have their car's oil change performed by someone else. Before your next oil service, use a permanent marker to write your initials on the side of the oil filter in your car. After the oil change is completed, which should include changing the oil filter, take a look and see if your initials are still there.

Another part of a good quality-assurance and quality-control program—and one that is often missed—is the level of documentation required. Entries in maintenance and equipment operational logs and documentation of work history and troubleshooting efforts are invaluable. A troubleshooting record, organized by symptoms, can save tremendous amounts of time when a piece of equipment fails. Again, if it's happened once, it can happen again.

If the time between equipment failures is long or the people who observed it last time are not immediately available, then it's effectively happening for the first time all over again. With a good equipment history/troubleshooting log, the personnel at hand can look up when the same symptoms were previously observed. The description(s) of the earlier troubleshooting efforts and identified causes may or may not be applicable to the current failure but can expedite returning the failed component to service.

Conclusion

Distributed energy systems, especially on-site cogeneration power plants, will need to employ creative energy storage systems to mitigate the increasing instability of the grid caused by the expanding renewable portfolios called for by state governments. As more and more renewable energy sources supply power to the grid, particularly wind and solar assets, the highly variable nature of these energy sources introduces both short-term risks (voltage and frequency variations) and long-term risks (outages). Federal, state, and local energy authorities, such as CAISO, are scrambling to find ways to balance their electrical distribution systems while complying with the mandates to include more renewable energy sources, yet the risks of system instability are felt directly by the individual consumers—including commercial and industrial facilities.

One way to manage the risk of increased instability in the local electrical grid is to use on-site power generation, and the most efficient way of doing that is by employing cogeneration. The use of energy storage systems, as part of a hybrid design, will be increasingly necessary to increase the reliability of such on-site cogeneration power plants. Risk mitigation strategies will vary from location to location around the country, but there will be a common objective: continuity of high-quality electrical power to the consumer.

There are advantages and disadvantages with any given energy storage solution to improve the reliability of an on-site cogeneration power plant, and individual facilities will need thorough planning and evaluation of the application of the available technologies. Many factors warrant consideration during the course of this evaluation, and the conventional use of the available technologies may or may not fit the given circumstances of that particular consumer. In particular, the use of currently available technology in unconventional ways, specifically battery and inverter systems, should be considered in on-site power plant applications.

Engine-generator sets (or combustion turbine generators) are typical components of on-site cogeneration power plants, and their operating characteristics are well understood. Using a battery system for energy storage with one or more inverters operating in parallel with the engine-driven generators presents some opportunities for increased reliability, demand charge mitigation, and peak shaving of the load profile, but it also presents some engineering challenges. The batteries, inverters, and inverter control software would be operating in a very different environment from their usual application in PV, and the power plant load sharing (and load shedding) control systems would need increased automation to account for all of the possible configurations and transient response.

Work is currently needed on normal generator operations, in parallel with the electrical utility; contingency operations if an engine trips off-line or a utility outage occurs; and the resulting electrical transients if the power plant shifts to island mode (microgrid). This book has described research and conceptual designs for such an energy storage system application in an on-site cogeneration power plant. People all over the nation are adopting this to obtain energy independence. They are realizing the savings and increased security of having the continuous power they need (to do whatever it is they do), rather than relying on their local utility.

Appendix A

Common Omissions and Failures

1. Failure to account for engine performance differences at different operating loads

The cost of the fuel consumed by a prime mover, whether an engine or a combustion turbine, is directly (and negatively) affected by the load under which the equipment is operating. Given that the two biggest drivers of the financial performance of a cogeneration power plant are the heat rate of the equipment (fuel costs in $/kWh) and the overhauls of that equipment, failure to account changes in these expenses as the units operate at varying loads will lead to overly optimistic financial forecasts.

2. Failure to account for seasonal and weather-related changes in system performance

Combustion-based prime movers will operate very differently in summer and winter conditions, and this difference in performance must be taken into account. The adequacy of the designed heat rejection systems and freezing protection, summer derating effects, and uninsulated steam piping becoming condensers are examples of seasonal and weather-related changes in system performance.

3. Failure to account for changes in load while operating in island mode

This is a huge—yet very common—mistake. The timing of the interactions among the control systems for the prime mover, plus the severity of the voltage and frequency transients, can have profound effects on the viability of the power plant to continue operations in island mode. Part of the problem is that this is very difficult to predict accurately. Often, power plants are built that *should* be able operate in island mode just fine. The largest electrical load is less than 30% of the capacity of the electrical generators, soft starts or variable-frequency drives (VFDs) are used on the largest motors, and the OEM's sales representative says their equipment is used for that purpose all the time.

In these cases, when the generating units fail to stay online during electrical transients, the devil is in the troubleshooting. I have seen a power plant where the culprit was determined to be the turbo lag in the engine's response to changes in electrical load, where that slight time delay was enough to exceed the timing limits of the protection relays. Sometimes the engineers hunting for a solution have get creative. Air from the local compressed air system can be used to keep turbos spinning at artificially high speeds, for example. In another case, a 200 kW battery-inverter uninterruptible power supply (UPS) system was used to provide a faster response time to the transient than the gensets could provide on their own, supplying just enough power to the main AC bus to allow the engines the time they needed to respond.

4. Failure to account for individual system integration

This is a problem when tying the equipment from the on-site cogeneration power plant into the customer's existing systems or when adding new equipment to an existing facility. The construction of electrical generators includes a *winding pitch*, such as 2/3 or 5/6, and this is an often ignored detail. When multiple generators are operated in parallel with each other, there can be problems if all of the generators are not of the same pitch (e.g., circulating currents on the common neutral). Also, if there are two or more generators operating in parallel that are of different sizes, the sharing of the available electrical load will not be equal, but instead will be in ratio to their respective size. Building an on-site cogeneration power plant from spare parts and used equipment in an à la carte fashion is almost always a recipe for failure.

If the customer is a food processor, there are NSF-rated heat exchangers to consider using, as well as a much more restricted list of acceptable chemical treatments for steam and water systems. The normal operating pressure of one system cannot be beyond the capacity of the customer's system it's supposed to interface with, among other considerations.

5. Failure to account for all constituent elements in water and exhaust

There are impurities in just about everything these days, including the air and water that are used in any on-site power plant. If the raw water supply or the city water supply is hard water, then a lot of problems can be minimized by using a softener system at the supply point. If the local air is polluted by the exhausts of other systems or processes occurring at the site or leakage from existing ammonia-based systems (or from anything else), then it must be addressed. The fouling of filters in reverse osmosis (RO) systems, for example, can be both an operational headache and an unexpected operating expense. Steam and water lines will get clogged with scale, as will heat transfer surfaces in boilers and HRSGs. Any contaminant vapor in the air that is inhaled by an engine or combustion turbine will alter the chemistry of the combustion process, sending the emissions out of specification. If the power plant is required to have continuous emissions monitoring, then even though there is nothing wrong with the equipment, all of the bursts of excessive emissions must be reported to the local air-quality management entity (usually, anyway).

Appendix B

General Questions to Ask a Prospective Energy Services Company

1. How many years have they been in business?
2. How many cogeneration power plant projects have they completed? How successful have they been?
3. How many years of experience in O&M do they have?
4. What is the O&M experience of the design engineer(s)?
5. What is their range of experience dealing with multiple equipment manufacturers?
6. What is their ability to provide financing? Are they willing to go open book?
7. If the ESCO will be providing O&M services, what is the experience level of their crew? What is the experience level of the crew's supervisor?
8. What spare-parts inventory do they recommend and why?
9. What is the nature and extent of any warranties on the equipment and power plant systems?
10. What is the time line of the project, including commissioning?
11. Are there plans to use the carbon dioxide and water vapor from the exhaust?
12. What references can they provide?

Appendix C

True-False Questions to Ask a Prospective Energy Services Company

1. Using the internal water pump of a stationary reciprocating engine as the only source of fluid flow for extensive and complex heat recovery flow paths is accepted as good engineering practice.
2. The operating environment of the equipment has no bearing on the design of the power plant.
3. The equipment nameplate data comprise all the information needed to accurately predict the operating performance of a power plant.
4. Performing reliability analyses, including studies of single-point safe elements, is a waste of time in power plant development and is therefore not a very smart idea.
5. Using duct burners in an HRSG located downstream of the NO_x and SO_x catalyst is never a problem for overall power plant emissions.
6. Operating stationary reciprocating engines or combustion turbines for extended periods of time at loads less than 30% of full power is never an O&M concern.
7. Having a robust quality-assurance/quality-control system in place during the construction phase of the power plant is not necessary.
8. Placing stationary reciprocating engines inside containers eliminates the possibility of any ventilation problems.
9. The failure of a $400 part would never force the shutdown of a $5 million power plant.
10. After a component or system of the power plant has failed, it is a waste of time to conduct a thorough investigation of the event.